ENERGY ECONOMICS:
A MODERN INTRODUCTION

ENERGY ECONOMICS:
A MODERN INTRODUCTION

by

Ferdinand E. Banks

Department of Economics
Uppsala University, Sweden

Kluwer Academic Publishers
Boston/Dordrecht/London

Distributors for North, Central and South America:
Kluwer Academic Publishers
101 Philip Drive
Assinippi Park
Norwell, Massachusetts 02061 USA
Telephone (781) 871-6600
Fax (781) 871-6528
E-Mail <kluwer@wkap.com>

Distributors for all other countries:
Kluwer Academic Publishers Group
Distribution Centre
Post Office Box 322
3300 AH Dordrecht, THE NETHERLANDS
Telephone 31 78 6392 392
Fax 31 78 6546 474
E-Mail <orderdept@wkap.nl>

 Electronic Services <http://www.wkap.nl>

Library of Congress Cataloging-in-Publication Data
Banks, Ferdinand E.
   Energy economics: a modern introduction / by Ferdinand E. Banks.
     p. cm.
   Includes bibliographical references and index.
    1. Energy industries. 2. Power resources. 3. Energy policy.

   HD9502.A2B348 2000
   333.79--dc21

                           99-052727

Printed on acid-free paper.

Printed in the United States of America

Contents

Preface

"Energy is the go of things", as James Clerk Maxwell pointed out. This simple truth was largely overlooked during the first 70 years of the 20th century, because in the industrial world most politicians, civil servants, and opinion makers were inclined to believe that virtually an infinite supply of reasonably priced energy would always be available, and so things would continue to 'go' in the manner to which many of their constituents and admirers had become accustomed. Similar opinions were held about fresh air, and water for consumption and agricultural uses. As a result, it was not until the last two decades of the century that serious courses in energy and environmental economics began to be offered at institutions of higher learning around the world.

This book is intended as a comprehensive introductory text and/or reference book for courses of this nature having to do with energy economics. (I have also attempted to make the book useful for self study.) As far as I know, there are no energy economics text or reference books on the level of this book in the English language. Needless to say, if I am wrong then I apologise to their authors; but right or wrong, I would like to see more energy economics books of all descriptions now. We cannot afford to have the same kind of mistakes made with energy policy that (in much of the world) are being made with e.g. employment policy. The stakes are simply too high. At the same time let me note that this is not a book on energy policy. Some policy matters may occasionally turn up in the exposition, but what I essentially want to do is to provide policy makers and their advisors with an elementary but thorough background in some mainstream theoretical concepts.

When I use the term 'level' above, I mean that if this book is used as a text, then it would be suitable for the first course in energy economics. It should be easily read by anyone with in interest in the subject who has completed a standard introductory course in economics. I have also tried to keep the mathematical (i.e. algebraic) content at that level.

A few readers, however, may be worried about the large number of exercises in the book. The only purpose of these exercises is to make readers thoroughly conversant with the materials in the book, and everybody should try a few: they are not intended to make excessive demands on the background or patience of the reader, and most of them are fairly easy, since their purpose is to build rather than undermine self-confidence. A few answers are given, especially for problems in the earlier chapters, but while I like answers for math problems, I am less enthusiastic on this matter where economics is concerned. There were hundreds of exercises involved in my courses on international finance, and no answers of any kind were provided. Instead I used a variant of the interactive teaching system once employed at the Ecole Polytechnique (Paris) by Joseph Fournier, and at the US Military Academy at West Point. Exercises were worked at the board by 3 or 4 students, who then discussed their work and answered my questions.

I would also like to emphasize that the oil market – mainly via oil derivatives (i.e. futures, options, and swaps) – has almost become an integral part of the financial market, and readers who master the materials in Chapters 3 and (especially) 6 on these subjects will be in an excellent position to continue their education in financial economics, as well as follow (and perhaps even participate in) the exciting developments that are just ahead in all energy derivatives markets.

In addition to being an elementary textbook, this book is intended to serve as a rapid refresher where energy topics are concerned. It is not a handbook, however. It was necessary to omit several important topics in order to ensure that most of the book – and preferably all of it – can be covered in a single course; and to make sure that all readers of this book obtain the basic knowledge they need in theoretical energy economics in a minimum of time. It may happen, however, that in a later edition of the book I will include such things as the sea transportation of oil and gas, hydroelectricity, and the outlook for alternative energy.

Finally, I would also like to apologise for the brashness of some of the text, but the journals to which I regularly contribute seem to appreciate this style, and perhaps it will make the present textbook easier to read.

Acknowledgements

I am grateful to a number of teachers and students for the help I have received over the years in preparing to write this book. First and foremost my students in international financial markets at Uppsala University, but also my teachers in the various departments of engineering at Illinois Institute of Technology (Chicago), and in economics at Roosevelt University in the same city, and at the University of Stockholm. I am also grateful to students and colleagues at Nanyang Technological University (Singapore); my students in energy economics at the University of Grenoble (France), the International Graduate School of the University of Stockholm, and Charles University (Prague); and my graduate students in mathematical economics at the University of New South Wales (Sydney) and at Uppsala. I should also like to thank Blackwell Publishers (UK), and note that throughout the years I have received invaluable help from the editors and referees of the OPEC Quarterly Review and the OPEC Bulletin. The Bulletin has always been filled with valuable statistics and information, while the Review is now emerging as one of the most valuable of the scholarly energy journals. I was able to make a decision as to what I would include in this book during my last visit to Australia and, more important, what I would not include. I therefore want to thank the Master, staff, and residents of Warrane College, the University of New South Wales (Sydney), as well as colleagues and friends at the University of New England (Armidale). Finally, I received enormous help from my colleagues at Uppsala University: Peter Gotlind, Åke Qvarfort, and Pebbe Selander; from my daughters Amelie and Madeleine, and my personal trainers Thomas Banks and Gunilla Banks; and especially from Javad Amin and his daughter Shadi Amin.

Chapter #1

Energy Economics: An Introductory Survey

Human beings are capable of mastering the natural world because they have developed cultures that are based, to one extent or another, on abstract reasoning. These cultures, together with such things as large amounts of inanimate energy, have worked to make life in cities such as Stockholm, Sydney, and Seattle a pleasure instead of a curse – at least much of the time.

This book is not about culture, but about energy; and the exposition will employ a small amount of abstract (or relatively abstract) reasoning derived from what is sometimes called 'economic science'. In this introductory chapter, however, the emphasis will be on descriptive materials, since one of my intentions is that *everyone* who reads this book will learn something about energy and economics, beginning with the former.

What about mathematics? These days economics tends to be inextricably associated with mathematics, but a considerable effort will be made in this book to separate most discussions involving more than simple algebra from the main text. *This is an introductory exposition*, and I personally make a point of avoiding highly mathematical presentations except on special occasions, and advise others to do the same. (This was also the advice of people like Albert Einstein and the Italian physicist and Nobel Prize winner Enrico Fermi – sometimes called 'The Pope', because he was never wrong.) The truth of the matter is that in economics a thorough knowledge of algebra, elementary calculus, and history is far more useful than a superficial knowledge of 'analysis' and set theory. By the same token, I would argue that the most important courses in economics are those at the intermediate level.

I also want to emphasize that I have written a book which, if used as a textbook, is primarily intended for a one semester (or term) course. There are, I admit, valid arguments in favor of a longer work, and I am in sympathy with some of them; but my experience in teaching economics and finance leads me to believe that most students are unable to absorb more than a book of this size in a single semester. Equally important, it is essential that persons who are working with energy topics should have access to a book that will refresh their knowledge of these topics in a minimum amount of time, and in addition provide them with a painless introduction to such important topics as derivatives and electricity economics. In a certain sense, this is a reference book as much as a textbook.

Anyone digesting the contents of this book can rest assured that he or she is capable of dealing with *all* the discussions about energy topics presented in such publications as *Time, Fortune, L'Expansion, The Economist*, etc, that are devoting an increasing amount of space to energy matters. They will also have no difficulty in handling much – though not all – of the specialized literature and, hopefully, discover how to present their own ideas about energy with a noticeable authority.

I have taken the liberty to scatter a number of exercises throughout this book. Many of these are extremely simple, but they should be done just the same, since among other things they are intended to build up the confidence of the reader.

1.1 Some Basic Ideas

The importance of energy in economic development is indisputable. Sweden, for example, is one of the most modern countries in the world, and its industrial and social progress can be largely attributed to its electricity intensive industry (and its up-to-date educational system). Similarly, at the global level, the consumption of energy in 1995 was more than 30% larger than in 1965, and predictions are that it will increase by an even faster rate before 2025.

There are many reasons for this acceleration in energy consumption, but among the most important are global population growth, the rapid transformation of certain parts of Asia into a consumer society of the North American/European variety; and steady industrial development in various other parts of the world – for example, Latin America. Take, for instance, the situation in Asia. The demand for automobiles and other durables in China, South Korea, Malaysia, etc. is going to have a drastic effect on the demand for and price of oil; and since in the long run other fossil fuels (e.g. gas and coal) are substitutes for oil, their prices will also have to rise, and perhaps by a great deal.

This ability to substitute for oil is often recognized in the market for natural gas by indexing its price to that of crude oil, where 'crude' refers to unprocessed oil: oil at the 'wellhead', as it is sometimes put. For example, on one contract, the price of natural gas (P_g) in dollars per million British Thermal Units (= $/MBtu) was given by

Equation (1.1). (Note that M signifies millions, and Btu British Thermal Units).

$$P_g = 3.65 \left[\frac{Average \quad price \quad of \quad 5 \quad crudes}{27.444} \right] \qquad (1.1)$$

In this expression 27.444 was the average dollar price of a barrel of crude oil (= $27.444/b) at the time this particular contract was signed, while 3.65 dollars per million British Thermal Units (=$3.65/MBtu) was the chosen base price for gas. The 5 crudes are predesignated, and not chosen at random by one or the other transactor. Interestingly enough, some contracts for coal are now being indexed to gas, and it happens that some are indexed to oil and gas. (The significance of the Btu will be explained in detail in the next section.)

As insinuated above, oil is the most flexible energy resource, and it will continue to be an all-around energy input in most of the developing countries. But in the industrial world, oil is mainly required to produce motor fuel. In such important usages as the generation of electricity, the major natural resource inputs just now are coal, uranium (for nuclear plants), and increasingly, natural gas. Although it is not generally recognized, electricity has become the second most prominent topic in energy economics since, worldwide, the rate of growth of electricity consumption is increasing faster than the rates of growth of the fossil fuels.

Some observers claim that the 21st century will feature a series of electricity crises. This may sound far-fetched until we hear that the World Energy Council (WEC) expects electricity investments to account for one-third of all energy investments in the period up to 2020. If this is true, then we are talking about trillions of dollars being spent for power plants, transmission lines etc.

It needs to be emphasized that electricity economics is not limited to studying the fuels that will be used to generate electricity, or the enormous problems that will be encountered in financing new capacity, but also the 'form' of the industry in every country. Is this industry going to consist of comparatively few very large production units, or very many generating stations that are fairly small, or what? Will the industry be heavily regulated, or left to the working of market forces, or something in between? There is also a matter of the environmental problems associated with the generation of electricity. These include disposal problems raised by nuclear waste, and various forms of pollution caused by fossil fuels. These problems can be solved, but it would be a very bad mistake to underestimate their seriousness.

It also needs to be thoroughly understood that regardless of the environmental laws passed by the industrial world, China and India cannot avoid using tremendous amounts of coal in facilities that are not equipped to suppress the pollution that is an integral part of burning coal; and everyone should be aware of the rapid increase in atmospheric concentrations of methane – which to a certain extent has its origin in the production, distribution, and burning of natural gas – and is now regarded as an important contributor to global warming via the 'Greenhouse Effect'.

Until a few years ago natural gas was highly praised as environmentally friendly, since when burned it releases less carbon than oil or coal, it contains hardly any sulfur, and its output of nitrogen oxides can be limited; but recently a number of investigators have flatly stated that gas represents an even greater environmental threat than oil or coal because of its methane content. This particular opinion is far from universally accepted however, and in addition there are copious sources of methane (to include rice paddies) that cannot be readily dispensed with.

Recently, in Sweden, a number of pro-nuclear advocates have asserted that the only way to avoid the environmental problems that result when electricity is generated using fossil fuels, is to use as much nuclear energy as possible. In addition they claim that 'safe' reactors, such as the Swedish PIUS and SECURE designs, will eliminate scenarios of the Three Mile Island and Chernobyl variety. Some people also say that in the long run it may be possible to develop reactors that will 'burn up' so much of a reactor's nuclear fuel that the storage of nuclear waste will no longer give rise to a 'Faustian dilemma'. (This is a dilemma where advantages in the short run are purchased at the cost of extreme disadvantages in the long run. The prototype here is Goethe's Doctor Faust, who sold his soul to the devil. Similarly, Thomas Mann's Doctor Faustus entered into what he thought was a highly promising agreement with the Prince of Darkness, only to have things turn out badly in the end.) A Nobel laureate in physics, Carlo Rubbia, says that he can build such a reactor.

The main difficulty with the new and alledgedly safe reactors is paying for full-scale versions of them, rather than relatively inexpensive laboratory or pilot models. This equipment is going to cost an enormous amount of money. So much, some people are saying, that they may never put in an appearance, since once development costs are factored into their unit costs, it will probably make them more expensive than alternative energy sources, although this is not certain. But even so, around the middle of the 21st century, a great deal of electricity will be needed, or desired, and if this non-hazardous gear is not available, then there will be a powerful temptation in some parts of the world to satisfy this need by a wholesale resort to plutonium fueled equipment. *Ceteris paribus*, this is not something to look forward to.

In the meantime the puzzle that overshadows all others has to do with the total supply of energy resources, both known and hypothetical, and to a certain extent with technological progress: will there be enough oil, gas and coal to maintain standards of living in the industrial world, and raise these standards in the developing countries, until science and technology can provide us with economical alternative sources of energy (such as solar and wind energy)? I certainly am an optimist, although my brand of optimism is tempered with realism. I also happen to believe that there is no such thing as a credible microeconomic theory of technological progress, and much of the talk about technology riding to our rescue if we paint ourselves into an energy corner should be heavily discounted.

A few numbers might be useful here. If we divide the amount of (known) oil reserves in the world by the annual global production of oil, which is generally slightly higher than oil consumption, then we get what some observers mistakenly call 'the length of life of oil reserves'. At the present time, this is taken to be about 41 years. As will be shown later in this book, however, this figure does not have a great deal of meaning, because it suggests that the present production (which is about 75 million barrels per day (= 75 Mb/d)) could be sustained for approximately 41 years. But, according to petroleum engineers, under normal circumstances oil deposits decline after about half of the reserves in the deposit are expended, and so current production can be sustained only if new discoveries match production. This was not taking place at the end of the 20th century.

Something else that is not taking place is a slowing down in the growth of the world's vehicle population. A few years ago Royal Dutch Shell estimated that the growth rate of private cars in the world was almost 12 percent/year, and recently the International Energy Agency (IEA) estimated that the number of vehicles on the road in the beginning of the 21st century will be greater than 700 million, which is an increase of approximately 30% in a decade. In addition, in the last months of the 20th century, the world car industry had an annual capacity of 60M cars/year. A theory now circulating among many observers is that the next oil price shock – if there is a 'shock' instead of a steady growth in the price of oil (and energy) at an uncomfortably high rate – will be directly attributable to the growth in private automobile ownership in Asia. It has also been estimated that globally the automobile sector is responsible for 25% of the carbon dioxide (CO_2) emissions from fossil fuel use, although technological progress should be capable of substantially lowering this figure. One hopes that this is true, because collectively the 700 million vehicles mentioned above will be emitting about 3.5 billion tonnes of CO_2 per year. Unfortunately, however, even after environmentally friendly automobiles begin to be constructed in large numbers, there is a large probability that – to begin with – the present variety will continue to be produced in even larger numbers, and thus aggregate CO_2 production will only decline very slowly.

Even if oil exploration was yielding satisfactory results on the whole, and the growth in car ownership and the average number of kilometers driven (per car) decreased, oil consumers could still be in trouble. Oil is not distributed uniformly over the globe. About three-fourths of present known reserves are in the Middle East, and this is also the region believed to be the richest in undiscovered reserves. In the aggregate of oil producing areas outside the Middle East, the nominal 'length of life of oil' is only about 18 years at the time this is being written, and it is falling – although fairly slowly. Many important geologists (and also the International Energy Agency of the OECD) claim that before 2010 oil production could turn down in the non-Middle East 'oil patch', although the former Soviet Union (FSU) might provide oil importers with some pleasant surprises: the Caspian area is repeatedly touted as a new 'Middle East', but this has yet to be proved. If these surprises are not forthcoming, then increased global dependence on Middle-East oil should assure the producers of that region that they will face a very favorable market situation. With world oil consumption well above 80 Mb/d by that time, it will not be necessary for these countries to pattern their behavior after rules formulated in Washington or Brussels.

This, however, does not have to be a disaster. As many prominent and wise persons in the Organization of Petroleum Exporting Countries (OPEC) have attempted to explain to their customers in the oil importing counties, there is actually much less petroleum in the crust of the earth than many persons think, and because of the structure of ownership and production that characterizes the world oil market, this market is not likely to function the way that markets functioned in the early chapters of the textbook that you used in your first course in economics. Simple marginal cost pricing may not be possible; and in the run-up to the oil market endgame, it would be a good thing if the people and governments of the oil importing countries took a more realistic attitude toward energy supply and demand, and the impact that even temporary oil (and energy shortages) could have on such things as macroeconomic performance and political stability.

If oil is destined to cause its buyers more than a few headaches, then what about natural gas, which is often labeled "the fuel of the future". The problem here is defining the 'future'. According to the World Energy Congress, global demand in 2020 will be twice that in 1995, reaching 4 trillion cubic meters (= 4 Tcm), which will be almost as much as the current gas reserves of the United States. I think that we need to be careful before making too much of an estimate of this nature, however, since in the US new technology has rejuvenated old fields, and allowed production to continue in fields previously believed uneconomic. Among these new technologies are 3-D seismic surveys, new drilling techniques, and new floating production systems that are the key to expanding deep water exploration and development. There are also very large gas reserves in the Middle East, both known and hypothetical, and it will not take any technological miracles to get these to market when the time comes. Much of this gas may eventually be liquified and sold as vehicle or aircraft fuel.

The gas being referred to above is 'conventional' natural gas. It seems likely that a great deal of gas will eventually be found in unconventional deposits: so called 'deep gas' that may be available if drilling can take place well below present maximum depths, or methane rich gas that might be located under or in the vicinity of large deposits of coal, etc. If this kind of gas can be exploited, then it might brighten the gas picture considerably, although by the time it becomes available, a great deal will be necessary. In addition, nobody knows how expensive very large supplies of deep gas will be, and the same is true for such things as 'heavy oil', and oil from shale. Note the emphasis on large supplies. Too many economists like to draw conclusions about the availability of unconventional resources from the performance of small-scale (or pilot) installations. In some situations this could turn out to be a serious mistake.

That brings us back to to coal. It is possible to make coal environmentally acceptable, and apparently at a cost that is low enough so that most rich countries may be able to burn increasing amounts of coal without it causing an unacceptable amount of pollution. Unfortunately, however, this observation might not be relevant (for economic reasons) for countries such as India and China. Given this situation, and since there is only one upper atmosphere, coal use in Europe and North America may have to be restrained for environmental reasons. To get an idea of what is at stake, the total climactic impact of rising greenhouse levels – calculated for presentation at the 1997 (Kyoto) meeting of the World Climate Conference – meant a doubling of preindustrial CO_2 concentrations by the year 2030, and a tripling (or more) by 2100. As for temperatures, the Hadley Center for Climate Prediction and Research (UK) has analyzed the global climate record for the 20th century in an effort to explain a rise of one degree Farenheit in the Earth's average surface temperature, and their calculations indicate that while before 1970 temperature increases could be explained by solar radiation and heat-trapping greenhouse gases, perhaps in equal proportion, after 1970 – when about half the century's warming took place – greenhouse gases were largely responsible. Policy prescriptions directed

8

toward a reduction in manmade influences on global warming include a call to reduce CO_2 emissions to about 30% of their 1997 levels in order to stabilize these emissions at double the 1997 value. The calculations mentioned above are not distinguished by a high degree of precision, because a great deal still needs to be known about the degree of natural (i.e. non-manmade) climate variability, but they are highly suggestive. Even in their present form they point to a potentially dangerous predicament, although according to Delia Villagrasa of Climate Action Network Europe, "This is not a problem where you have to invent solutions. We have the solutions. What is needed is the political will to make it happen."

The quantity of coal in the crust of the earth is huge, as Figure 1.1 indicates. Moreover, unlike oil and gas, coal is found in many parts of the world, although in former large producing countries such as the UK and Germany, its output is falling, and voices are heard calling for the closing down of installations exploiting all except the richest deposits. Much coal is consumed in the region where it is produced, although many industrial countries are large importers of coal from South Africa, Australia, etc. The United States is a major coal exporter, while in the last two decades, the largest growth in output and exports has take place in Australia.

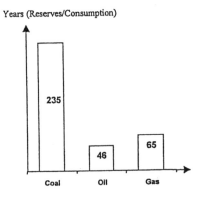

Years (Reserves/Consumption)

RESERVES DISTRIBUTION (By area) in 1996

	Gas[a]	Coal[b]	Oil[c]
North America	8.4	250.0	86.6
South America	5.7	10.2	78.9
Europe	5.5	157.0	17.7
Africa + Middle east	54.6	61.9	732.6
FSU	56.0	241.0	57.0
Asia + Australia	9.5	311.5	44.1

a: In Trillion Cubic Meters (Tm^3)
b: In Thousand Million Tonnes (Gt) (includes coking coal)
c: Thousand Million Barrels

Figure 1.1

From this diagram it appears that coal is the ultimate *backstop* resource: it not only can be used for heating and the generation of electricity, but if necessary it can be used as an input to produce motor fuel and lubricating oil, though of a poor quality unless this technology is upgraded, or unless this oil can be further processed, which raises its cost. Thus, should the oil and gas taps begin to run dry, it is theoretically possible to turn to coal.

During the latter stages of World War II, Germany produced large amounts of oil from coal. Sasol Ltd, of Johannesburg South Africa, improved and utilized this technology. Sasol now produces diesel fuel from gasified coal (i.e. 'syngas'), and is attempting to gain access to gas supplies elsewhere in Africa in order to produce another variety of 'synthetic crude'. The interesting thing for developing countries is that if gas from smaller fields can be economically transformed to synthetic crude, and thus can be loaded on a tanker or put into an existing pipeline, it will often be much more profitable to exploit than if it had remained in gaseous form.

Even if the coal to oil process can be greatly improved, some experts seem to think that we are not many decades away from the unpleasant point in time when reserves of high quality coal will start becoming scarce relative to demand. The thing to note here is that, as with oil, even a low growth rate of consumption will eventually lead to a high absolute level of consumption, at which time a low rate of growth of consumption will cause large absolute increments in consumption.

When the total oil consumption in the world was 18,250 Mb/year (= 50 Mb/d) a growth rate of 1.75 %/y did not cause any undue pressure on the price, since it was possible for supply to expand at such a rate as to accommodate the increased demand. But when just under 27,000 Mb/y are being consumed, as at present, a growth rate of consumption of e.g. 1.75%/y – calling for an annual increase in supply of 27,000 x 0.0175 = 472 Mb/y in order to keep pace with the growth in demand – becomes progressively more difficult to bring about, given certain geological, economic, and political realities that will be taken up in Chapter 3.

There are also some grave misunderstandings about the effect of low growth rates of consumption on even a very high level of reserve accumulation. Let us consider the following approximate formula. (Exact for continuous compounding).

$$T_e = \frac{1}{g} Ln\left(\frac{g\overline{X}}{X_0} + 1\right) \qquad (g \neq 0) \qquad (1.2)$$

In this expression T_e is the time to exhaustion of an amount of a resource, $\overline{X}$, having a growth rate of consumption of g, with X_0 the initial amount consumed. Ln signifies the natural logarithm – in this case the Ln of the value in parenthesis. Now lets assume that g = 2.5%/y, X_0 =100, and $\overline{X}$ = 23,500. We thus obtain:

$$T_e = \frac{1}{0.025} Ln\left[\frac{0.025 x 23,500}{100} + 1\right] = 77.1 \ years$$

If we use the concept introduced earlier of the 'length of life of a resource', then the (static) length of life of world coal reserves is shown in Figure 1.1. But if we look at this matter 'dynamically', with a modest growth rate of consumption of the resource of 2.5%/y, then the length of life is reduced to 77.1 years.

Next let us assume that we double the availability of the resource, going from 23,500 to 47,000 units. Our calculation for T_e using (1.2) is thus:

$$T_e = \frac{1}{0.025} Ln\left[\frac{0.025x47000}{100}+1\right] = 101.82 \quad years$$

Note what happened! We increased the amount of the resource by 100%, but the (dynamic) length of life (T_e) was only increased by (101.82–77.1)/77.1 = 0.32 =0.32 x 100% = 32%.

For reasons that will become obvious in chapter 3, the exercise carried out above does not strictly apply to oil or gas, although it is fully applicable to coal. Even so, M. King Hubbert had something like this in mind when he pointed out that even if oil exploration were able to uncover a few giant or super-giant oil fields, in the long run these new discoveries would not delay the turning down of oil production by more than a few years. This observation also applies to a comparatively small increase in the percent of oil that can be recovered from a deposit, and at the end of the 20th century, only 'comparatively small' increases seemed likely. Clearly, the 21st century is going to require some profound technological achievements as well as some deep and imaginative thinking on the part of our political masters if we are to be able to keep the lights on and the motors running the way we did during the 20th century.

1.2 Units and heat equivalents

This section is meant to give readers a knack for working with the units employed in energy topics. At the same time it should be understood that the section is incomplete, and deliberately so, since additional materials dealing with these matters will be introduced later as they are needed. A word to the wise might also be useful. Many students never learn as much macro economics as they could learn if they spent more time studying national accounting and flow-of-funds analysis. Similarly, a fluency with energy units is essential for a complete understanding of energy economics.

We start with an important definition. *Primary energy* is energy obtained from the direct burning of coal, gas, and oil, as well as electricity having a hydro or nuclear origin. Electricity obtained from burning fossil fuels is a secondary energy source, and the same is true of electricity produced from such things as 'town gas' and coke.

We also need to know that one metric ton (= 1 tonne = 1t) equals 2,205 pounds, and that 2.2 pounds is equal to one kilogram. (Similarly, 1 inch = 2.54 centimeters). In everyday life the usual ton is the short ton, or simply ton, which equals 2,000 pounds. Thus, 1t = 1.1023 tons. It does not hurt to know that there is also a long ton, and this is 2,240 pounds.

Now we go to some symbols used for energy units and conversion factors. Observe the following tableau:

Table 1.1

Prefix	Power	Symbol	Meaning	Example
Kilo	k	10^3	thousand	kW (kilowatt)
Mega	M	10^6	million	MW (megawatt)
Giga	G	10^9	billion	GW (gigawatt)
Tera	T	10^{12}	trillion	TW (terawatt)
Peta	P	1015	Thousand-trillion	PW (petawatt)
Exa	E	10^{18}	Million-trillion	EW (exawatt)

In the most elementary, yet most comprehensive sense, energy can be defined as anything that makes it possible to do work – i.e. bring about movement against resistance. Energy takes many forms, and one of its most interesting characteristics is that all aspects of motion, all physical processes, involve to one degree or another the conversion of energy from one state to another. For example, the chemical energy that is found in coal can be converted to active heat, which in combination with water will generate steam in a boiler. This steam can then be used to drive a turbine which, in turn, rotates the shaft of an electric generator, and thus produces electricity. Note also that the rotating shaft implies the ability to do physical work. For instance, it could be used to turn a merry-go-round, or a water wheel, or even to pull a cable car up Coogee Bay Road in Sydney's Eastern suburb (although, admittedly, these are probably very uneconomical applications). Analogously, the chemical energy in food can be transformed into mechanical energy – i.e. the ability to do physical work – or, if a person is so inclined, the ability to do mental work. Figure 1.2 displays some of these concepts, and also presents some information that will be useful later. (The reason the year 1985 was chosen will be obvious by the end of this section).

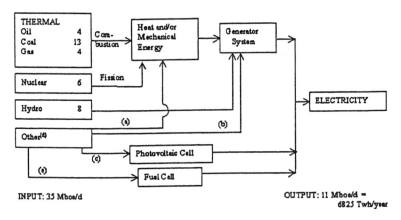

INPUT: 35 Mboe/d

OUTPUT: 11 Mboe/d =
6825 Twh/year

(a) Solar-Thermal: Biomass, Geothermal, Solar
(b) Wind, Tidal, Wave
(c) Solar
(d) Negligible
(e) Hydrogen-Oxygen

Source: Shell Briefing Service, 1986

Figure 1.2

As indicated in figure 1.2, the daily input of energy in electricity generation in 1985 averaged 35 million barrels of oil (equivalent), which can be written 35 Mboe/d, while the output is only 11 Mboe/d. This means a production and transmission efficiency of approximately 31.5 percent: slightly more than two-thirds of the energy in the inputs was lost before it reached the final consumers. Immediately below, readers will be given an initiation into the important technique of calculating 'equivalents', but note, considerable 'rounding' has taken place, and this may cause certain slight numerical discrepancies.

Let us begin this project by examining the significance of *power* and *energy*. Power is the time rate at which energy is converted. The unit of power is the International System of Units (SI) is the watt, although the watt is so small that the kilowatt (= 1,000 watts) is usually employed. We shall be using this unit of measurement a great deal in this book, and it is just as well that we put it into its proper perspective now. A pound of coal (= 0.4545 kilograms) contains on average 12,500 British Thermal Units (Btu) of heat energy. By definition, a Btu is the quantity of energy in heat form needed to raise the temperature of one pound of water (which is approximately one pint) by one degree Fahrenheit. Since 1t is equal to 2,205 pounds, it contains (on the average) 12,500 x 2,205 = 27,563,000 Btu, which will be rounded off in the calculations immediately following to 27.6 Mbtu. (Remember: 1t = 1 tonne = 1 Metric ton. You should also take particular notice of the use of the expression *average*.)

Now let us introduce two light bulbs. One of these produces a great deal of illumination, and has power rating of 500 watts; while the other is considerably weaker, and has a rating of only 50 watts. If the heat energy in coal is totally and perfectly transformed into electrical energy (i.e. with 100 percent efficiency), then 3,412 Btu are required to generate a kilowatt-hour (kWh) of electrical energy (where the kWh is the unit in which electrical *energy*, as distinguished from *power*, is measured). The power rating of the bulbs, 500 and 50 watts respectively informs us of the rate at which the energy potential of the coal is consumed; and so if the 27,600,000 Btu in a tonne of coal is transformed into electricity in a perfect system, then it could provide exactly 27,600,000/3412 = 8,090 kilowatt-hours of electrical energy. In other words, in a perfect system the stronger of the two bulbs, which consumes power at the *rate* of 500 watts (= 0.5 kW) could function for 8090/0.5 = 16,180 hours. The other bulb would require 8,090/0.05 = 161,800 hours to consume a tonne of coal. In reality, the efficiency with which fossil fuel can be converted to electrical energy is well under 100 percent. An efficiency of about 32 percent seems typical for much of the industrial world, and so on the average it would require 10,662 (= 3,412/0.32) Btu to obtain a kWh of electrical energy. This can be put another way: 1 kWh(e) = 3.12 kWh (fossil fuel). The UN and OECD also calculate e.g. that 1 kWh(e) = 2.6 kWh (oil).

It was stated above that a tonne of coal has an (average) energy content of 27,600,000 Btu. Similarly, a barrel (b) of oil has an (average) energy content of 5,800,000 Btu. It is averages of this sort which, in Figure 1.2, permitted us to discuss coal in terms of an oil equivalent. In Figure 1.2 we have 11 million barrels of oil equivalent. For example, 1t of coal is equal to 27,600,000/5,800,000 = 4.75 barrels of oil equivalent. In Figure 1.2 we have 11million barrels of oil equivalent per day (= 11 Mboe/d) output, which is 11 $\times 10^6 \times 5.8 \times 10^6 = 63.8 \times 10^{12}$ Btu. This can be converted to kilowatt-hours (= $63.8 \times 10^{12}/3412 = 1.869 \times 10^{10}$ kWh/d). Because these figures tend to be quite large, another unit – the Terawatt-hour (TWh) – might be more useful, where 1 TWh = 1 trillion watt-hours. Accordingly, the above figure becomes 18.69 TWh/d, or 6,825 TWh/y as shown in Figure 1.2. By the same token we could have used megawatt-hours (MWh) or gigawatt-hours (GWh) if these had caught our fancy.

As simple as all this seems, many readers may feel uneasy. While electrical power is defined as a 'rate', it is not always explicitly associated with a time dimension: for instance, the 'rating' of a power station is likely to be in megawatts. However in the example with the bulbs we saw that a large bulb 'milked' the ton of coal of its energy potential more rapidly than a small bulb, which trenchantly suggests that the dimension for power is energy per unit of time. As it happens, a watt is one joule per second (which is easily recognized as a rate) or 3,600 joules per hour; and since 1,055 joules is one Btu, one watt is 3.412 Btu/hour (which is even more easily recognized as a rate by those of us accustomed to working with the Btu.) Observe that 1 kW = 1,000 J/second, where J signifies joules.

This example can be extended by treating power as analogous to velocity, and energy as analogous to distance. The total distance traveled is velocity (kilometers/hour) multiplied by time (hours), just as the total energy converted is power (J/time or kilowatts) multiplied by time. The power rating of the vehicle determines (*ceteris paribus*) its speed, while the power rating of the bulb determines its illumination. In both cases, the larger the power rating, the more *work* can be done, where the units in which work is measured can be made the same of power. Put more succinctly, the major uses of energy are for the production of work or heat: the flow of energy is called work when it exerts a force, and heat when this is not the case. Work and heat are alternative modes for the flow of energy. Something else that we are undoubtedly aware of is that the more work the bulb or car can do, the higher its acquisition cost, and often the higher its operating cost.

Let's take this discussion a step further. The fuel in the tank of a car may generate 10 million Btu (= 10 MBtu) during an hour of driving. A portion of this energy – for example 3.5 MBtu – might be transformed into work in the form of rotating a shaft that turns the wheels of a vehicle. The rest of the energy is discharged as heat into the air (or, perhaps, into cooling water). Fuel efficiency in this example is thus only 35%, which is the percentage of the fuel that is actually transformed into *useful* work. Even more sadly, as the temperature of the 'non-useful' work falls, we are losing forever its availability to do work: its unavailability is increasing. This is what *entropy* is all about: the permanent degradation of energy.

One further item needs to be taken up before continuing. In the above discussion, I have used the Btu as the basic unit for heat energy. This is very often done in both the United States and Australia, with the 'therm' (= 1,000,000 Btu) being an important unit in Australia, and the Quad (= 10^{15} Btu) occupying a similar position in the United States. In much of the world the basic unit is the *joule* (J), and it seems that in serious scientific usage the joule is preferred everywhere. Fortunately, it is easy to go from one unit to the other. 1 billion joules (= 1 GJ) is equal to 947.8 x 10^3 Btu or, conversely, 1Btu = 1,055 J. Finally, 1 kilocalorie (kcal) = 3.968 Btu, and so 1 Joule = 2.4 x 10^{-4} kilocalories.

In the previous discussion the expression 'efficiency' was used when discussing the number of Btu necessary to obtain one kWh in an actual system. In practice the term *heat rate* is sometimes used, where heat rate can be defined as 3,412 (Btu) divided by the efficiency of a system or an installation. Knowing the cost of the fuel, and the heat rate, will permit calculation of the cost of fuel per kilowatt hour.

Consider now a situation where the cost of coal at the generating station is 36.3 dollars/tonne, which was the average landed price of *steam* (or thermal or steaming) coal at Rotterdam in the second half of 1985. (Somewhere toward the end of the first decade of the 21st century it is expected to be from 7 to 10 dollars more expensive). Again assuming 12,500 Btu/pound as the average energy content of coal, we get for a tonne of coal 12,500 x 2,205 = 27,562,500 ≈ 27,600,000 Btu. Taking the efficiency of transformation to be 32.5% (=0.325), the fuel cost of electricity can be expressed as:

$$\frac{36.3\left(\dfrac{Dollars}{Tonne}\right)x100\dfrac{Cents}{Dollar}}{27,600,000/3412/0.3250} = 1.381 \quad Cents/kWh$$

This figure is turned into *mills* by multiplying by 10. (Mills is a unit that is sometimes used.) We thus get 1.381 cents/kWh = 13.81 mills/kWh. It is also enlightening to check the units by employing the following dimensional scheme.

$$\frac{\dfrac{dollars}{tonne}x\dfrac{Cents}{Dollar}}{Btu/Tonne/Btu/kWh} = \frac{\dfrac{Dollars}{Tonne}x\dfrac{Cents}{Dollar}}{\dfrac{Btu}{Tonne}/\dfrac{Btu}{kWh}} = \frac{Cents}{kWh}$$

The reader should observe that in calculating Btu/kWh, the efficiency of conversion was used. If this is 32.5%, then Btu/kWh becomes 3,412/0.325 = 10,498, which has been called the *heat rate*.

1.2.1 Exercises

1. You have just started this book. By the time you finish, try to find some up-to-date information that can be used in Figure 1.2. For example, see if you can find a recent Shell Briefing Service publication with the latest figures that are applicable to this diagram. While doing that, solve the following. The daily output of the sun is 3 x 10^{32} joules. How many tonnes of coal is this equivalent to?
2. In the calculation directly above, convert Btu to Joules, and then compute the final cost of electricity in dollars/kWh! (i.e. the 'formula' for cents/kWh.)

3. Suppose that heating oil used to generate electricity costs \$25/b, which means that it costs \$183.2 /t, if we assume that – as with crude oil – we have 1t = 7.33b. What, then, is the fuel cost of electricity generated with oil (instead of coal as in above example), but assuming a conversion efficiency of 35%? In comparing this result with the calculation in the text, what can you say about the cost of generating electricity with oil compared to coal?

1.3 More Basic Ideas

In teaching international finance, I enjoy insisting that the subject has a rhythm to it. That rhythm can be acquired by the absolute mastery of a small amount of algebra, together with a thoroughgoing insight into elementary financial theory and history. Exactly the same thing is true of energy economics. There are a fairly small number of facts that we need to have at our fingertips all the time, while simultaneously recognizing that certain widely held ideas are completely without foundation.

Economics is an observational science, so let us look at the situation in the United States once more. What happened in that country is that a crucial statistics called the reserve-production ratio – the ratio of oil reserves to the production of oil, referred to a given period – fell below 10, and continued to decrease until it was close to 9. At that point, for very simple reasons that will be explained in Chapter 3, an extremely large decline in oil output took place. (Please note, however, that the production of oil in the 'lower 48' began declining about 1970, although there was a tremendous amount of producible oil in the ground at that time, and a similar observation applies to 1985, when oil production in the (entire) US began to fall.) As noted earlier, the same thing should take place – sooner or later – in the main oil producing countries outside OPEC or, for that matter, countries outside the Middle East; and then the oil output of this group of countries will also tumble. (This is a simple example of Einstein's equivalence principle: if two phenomena display equivalent effects, then they must be manifestations of the same fundamental laws.) That puts the Middle East oil producing countries in the position of having to supply a much larger share of world oil, which is a service that they are unlikely to be willing to provide unless the oil price rises by a healthy amount.

Something that is essential to understand here, and which applies to other fossil fuels, and perhaps uranium, is that when actual production gets close to the maximum output that can be produced at a given time, or full-capacity output, producers tend to raise prices. This behavior can be deduced from the production theory you learned in the first course in economics: when cost curves are U-shaped, high capacity utilization will result in high costs. At the most elementary level of abstraction we might posit a 'price equation' that appears to describe price response. This equation looks as follows

$$P_t = P_{t-1} + \phi(C_{ut} - \bar{C}_u) \tag{1.3}$$

P_t symbolizes price in the present period, P_{t-1} is the price in the previous period, $\bar{C}_u$ the desired level of capacity utilization, C_{ut} the actual level of capacity utilization in period t, and ϕ is a positive constant ($\phi > 0$). When $C_{ut} = \bar{C}_u$, or the actual capacity utilization is equal to the desired capacity utilization, the price is constant: $P_t = P_{t-1}$. But when, for example, $C_{ut} > \bar{C}_u$, then $P_t > P_{t-1}$: the price increases. This simple expression has occasionally been fairly successful in capturing OPEC's price behavior since 1975. If it continues to be reasonably valid, then the world oil market might be treated to a great deal of excitement before the end of the first decade of the 21st century, since actual OPEC capacity utilization moved from approximately 80% in 1990 to about 94% in 1996.

Oil will always be an important topic on the international conference circuit, but of late it has had to share the spotlight with the deregulation and, where relevant, privatization of the electricity and natural gas sectors.

The merits and demerits of electricity and gas regulation have been discussed for decades, but in the electricity sector it became a raging controversy in the late 1980s with the development of highly efficient *combined cycle* gas turbines. What happens here is that the hot exhaust gases resulting from the conventional power generation are confined in such a way that they can drive another turbine which generates more electricity. The efficiency of some of these plants is well above 40%, which happens to be very high, and according to some engineers, combined cycle equipment operating with extremely high turbine temperatures may eventually permit efficiencies in the vicinity of 50%. Although very large installations featuring gas turbines operating in tandem are employed in e.g. Japan, this materiel is supposed to be especially suited for small and medium sized electricity generating facilities.

If this is so, then instead of having a huge power station, functioning as a monopoly for a particular district, it might be possible to have a number of smaller installations whose deportment is similar to the production units in the model of perfect competition that you were introduced to early in your introductory economics course.

About the same time, in the United States, the demand for electricity slowed down. In this situation, it no longer seemed wise to build huge power plants whose most efficient operating capacity would not be reached for many years, and which would be uneconomical until that time unless given special treatment by the authorities. This is the origin of the so-called 'stranded investment' or 'stranded costs' dilemma, where this expression refers to those investments made by firms in the regulated sector which are expected to become uneconomical when competition is introduced – assuming, of course, that electricity rates fall. Major changes are expected to be delayed until the financial commitments of e.g. bondholders in these *utilities* – that is, monopolies that were regulated in order to keep them from earning profits that were considered socially inadmissible – could be safeguarded. Once this is done, an all-out effort would be made to decentralize and, where relevant, privatize the industry. This entails speeding up deregulation; unbundling – which means that various technical functions (e.g. generation, transmission, etc) would be separated, with separate ownership; and, where possible, electricity would be sold on spot markets instead of via long-term contracts. *Derivatives markets* – and particularly futures – would be introduced on the largest possible scale in order to facilitate guarding against the uncertainty that, earlier, had been dealt with through the straight-forward utilization of long-term contracts.

Many of these matters will be considered later in this book, however it is just as well to point out now that a decade after it was pronounced that the US suffered from a pronounced surplus in generating capacity, power shortages began to appear in some parts of the country. The simple reason for this was under-investment over a number of years, and as a result, just before the turn of the century, the US share of world orders for new power stations was set to quadruple. In fact, in the first decade of the 21st century, about \$80G were scheduled to be invested. As to be expected, however, during the same period Asia should account for at least twice as much additional generating capacity as North America.

Now that we know something about units, it might be beneficial to take a look at the best known renewable energies. Waterpower dominates here, with about 710,000 MW. It provides very inexpensive power, and may amount to as much as 2.5% of the global energy supply. A larger share is possible, but some observers say only at the cost of extensive ecological damage. By way of contrast, geothermal energy is generally considered very attractive, but there is not enough of it: in the year 2000 only about 700MW(e) and 11,000 MW(heat). It is more expensive than e.g. coal, but there is no pollution associated with it. Possibilities for expansion are, unfortunately, probably limited.

At the turn of the century there is about as much wind power [7000 MW(e)] as geothermal energy, but a great deal is apparently expected of this energy source. The cost of energy from wind is, on the average, somewhat above that of coal; but the absence of atmospheric pollution makes wind a very appealing alternative to fossil fuels. Great things are also eventually expected for solar-cell technologies (700+ MW at present), and solar power plants (700+ MW), but these are still comparatively expensive, and the amounts already in place are very modest. The really important breakthrough will come when better means are found for economically storing *large amounts* of wind and solar generated electricity. Whoever solves this brain twister will deserve the Nobel Prize that he or she should be awarded on the spot.

The production of electricity (and perhaps heat) from alternative energy sources is extremely important, but equally important is the development of vehicles that do less damage to the environment. Many environmentalists believe that these vehicles already exist, and they do, but they will not be purchased in large numbers until they become more attractive economically. The latest news from the scientists and engineers who are trying to design such vehicles is that they still have a great deal of very complicated work to carry out before a solution is found, however technologies involving methanol and fuel cells (see glossary) appear to be the direction in which present research is heading.

1.4   Conclusions

Hopefully, most readers have begun to grasp the main issues in energy economics, and how difficult it may be to solve energy related problems in the long run. No, we probably will not run out of oil, but we could be priced out of oil, especially if forecasts of the amounts that will be consumed in the coming decade or two hold. And what about the prediction by the WEC that we are not more than a few decades from the date when the annual world consumption of natural gas is as large as the presently known gas reserves of the United States? There is a great deal of gas in the crust of the earth, but even so this kind of news deserves to be taken very seriously by everyone looking forward to a relatively comfortable and stress-free life, and who desire the same thing for their relatives and descendents.

For example, weak oil (and energy) prices in the late 1990s were a key factor in reducing world inflation – especially in the US – which means that they played an important part in holding down interest rates. There has been a firm link between unemployment in the US and the oil price since the 1950s, and in a later period, only a slightly weaker relationship between energy prices and corporate profits. An increase in energy prices is capable of eroding corporate profits – directly or indirectly; and if the real return on capital does not decline, real wages must adjust: higher unemployment has often been an important factor in bringing this adjustment about. (You can also remember that in our Post-Keynesian world, higher inflation is transformed into higher interest rates.)

One of the most disturbing things about economics, and in particular energy economics, is the systematic playing down of such things as indivisiblities, increasing returns to scale, externalities (e.g. environmental spillovers in production and envy in consumption), incomplete financial markets, and uncertainty. But these things dominate in the real world, and especially uncertainty. Perhaps more than anything else, it is uncertainty that keeps economics from being the science that some people think that it is. Not only uncertainty about costs and benefits, but also about the behavior and tastes of 'the human factor'.

It needs to be made particularly clear that one of the main things causing economies of scale in production is the universality of indivisibilities. Pipelines, transmission lines, and to a considerable extent power plants and the compressors used to push gas through pipelines tend to only be available in specific sizes, and their economic usefulness manifests itself only when the scale of operation is large. In addition, as noted by Scarf (1994), the assumption of constant returns to scale that we teachers of economics are so fond of making, can lead to conclusions about the organizing of an industrial society that are outrageously naïve.

Very little is said in this book about unconventional (i.e. alternative) energy sources, although by the middle of the 21st century they must be available, and available in *very large* quantities. (Note the emphasis on very large.) The end of the Cold War has released thousands of world class scientists for the research needed to turn the final corner where wind and solar energy, among others, are concerned. They need to be put to work on this project immediately, and given all the political backing they might need. Tomorrow may be too late.

1.4.1 Exercises

1. An interesting formula for the pricing of gas has been used on some contracts involving Indonesian gas. It was:

$$P_{gas} = P_{oil}\left(0.5\frac{A}{11} + 0.5\frac{W}{135}\right)$$

A is the price of Indonesian crude oil (which was $11/b in the base period), while W is an index constructed from the price of some commodities imported by Indonesia. Discuss this formula, which features a certain concept that is favored by many developing countries. What is this concept?

2. Using equation (1.2) and the numbers in the example following this equation, what effect would a growth rate of 5% have on T_e?

3. Study equation (1.3), and remember your first course in economics. If this equation was used to set the price of oil by OPEC, what would happen with production elsewhere? Could this equation, or something like it, apply to any firm?
4. 1 Btu raises the temperature of one pound of water 1 degree Fahrenheit (F). How many degrees Celsius (C) will one kcal raise 1 kilogram of water if we have $C = 5(F - 32)/9$, and so $\Delta C / \Delta F = 5/9$?

1.4.2 Answer Quickly and Briefly

1. In equation (1.3), assume that $C_{ut} < \bar{C}_u$. Describe the price dynamics!
2. Here are some questions involving macroeconomics! Would rising oil prices raise the inflation rate? Why would higher energy prices erode corporate profits? Why would a higher inflation rate lead to higher interest rates?
3. CIPEC, an organization formed by copper producers, never succeeded in becoming an OPEC. Why might this be?
4. Building and operating oil tankers takes good nerves, as Aristotle Onassis liked to point out (after he became very, very rich). What are some of the problems that shipowners like Mr Onassis must face?
5. Economists feel that science, technology, and the market will see to it that we have all the energy we need, while geologists are not so sure. How do you feel about this?

Chapter #2

Discounting And Capital Values

These days, almost every economics textbook contains materials on discounting. The problem is that many of them do not contain enough, and certain very important details are omitted. For example, they do not spend enough time on the important theoretical concepts that serve as a background to standard discounting techniques. On the other hand, there is no denying that many energy assets consistently trade at market values that are very different from than those that can be calculated using these techniques, nor that the *ad-hoc* adjustments that investors inevitably apply to discounting exercises originate well outside the scope of mainstream theory.

For more on this topic the reader is referred to Davis (1996), although while going through the present exposition, an observation of Professor Darwin C. Hall should be kept in mind: "Theory improves over time, given the energetic attempts to apply it and the repeated discoveries of serious shortcomings."

Despite appearances, the presentation in this chapter does *not* involve mathematics above the level of secondary school algebra; and where this algebra is concerned, only one comparatively sophisticated operation appears. It is taken up early in the first section. Chapter 8 examines some related materials, while Chapter 9 discusses, on an elementary level, some elementary mechanics of real option theory, in which uncertainty and irreversibility are factored into decision making.

2.1 First Principles

In working through this chapter, readers will be dealing with a simple procedure, which begins with obtaining the *net present value* (NPV) of an investment. This involves calculating the present value (PV) of the (expected) revenues that an investment will generate over time, and subtracting from this the present value of *all* the (expected) expenditures or costs required to implement the project. That calculation gives us the NPV. Now, if the NPV is greater than zero (NPV > 0), then the investment is judged viable – although in Chapter 9 some questions are raised as to the overall adequacy of this approach. For the time being, however, those questions will be overlooked – just as they are mostly overlooked in many reputable books on corporate finance.

Suppose that you put a thousand dollars in a bank, and the interest rate is 10% (= 0.10). At the end of the year you have $1000(1 + 0.10) = 1,100$ dollars, if the compounding is done once a year (at the end of the year). Now put the money in for 2 years. If compounding again takes place once a year, at the end of the year, you would have at the end of the 2 years $1000(1 + 0.1)(1+ 0.1) = 1000(1+0.1)^2 = 1,210$ dollars. Recognize something else here! In the first example, 1,000 dollars is the *present value* (= PV) of 1,100 dollars – or, by a similar token, 1,100 is the *final value* (= F) of 1,000 dollars. In the second example 1,000 is the PV of 1,210 received at the end of the second year. That is, $PV = 1210/(1 + 0.1)^2 = \$1,000$. This operation should be understood perfectly before you continue.

We can continue by asking what is the present value of $5,000, received at the end of 4 years, if the interest rate is 12%, and compounding takes place once a year (at the end of the year)? The answer is $5000/(1 + 0.12)^4 = \$2,542$. What is the future value of $1,200 after 10 years, if the rate of interest (= r) is 6%? This is $1200(1 + r)^{10} = 1200(1+0.06)^{10} = \$2,149$. The reader should perform these exercises with an interest rate of 10%.

Now suppose that we put $1,000 in a bank, and compounding takes place twice a year: first at the end of 6 months, and then again at the end of the year. If the rate of interest is 10%, F at the end of the year is $1000(1 + 0.10/2)^2 = 1000(1 + 0.05)^2 = \$1,102.5$. We thus suspect that as we increase the number of compoundings, F increases; but before we proceed to the ultimate with an 'infinite number' of compoundings per year, let us take a two-year situation with 3 compoundings per year (i.e. every 4 months), and r = 10% (0.10). Now we have $F = 1000(1 + (0.10/3))^{3 \times 2} = 1000(1 + (.10/3))^6 = \$1,217.4$. Compare this with F at the end of 2 years, but with only one compounding per year, and once again you will see the result of multiple compoundings. (Observe: A term that sometimes appears is the 'effective annual rate of interest', or r_e. With a one year situation (T = 1), r = 10%, and two compoundings per year, the effective rate of interest is $r_e = (1 + 0.10/2)^2 - 1 = 0.1025 = 10.25\%$.)

Dismissing the expression present value for the time being, and considering an amount of money at time 'zero' equal to P, then with an interest rate *r* we can write:

1 year; 1 compounding /year	$F = P(1+r)$
T years; 1 compounding/year:	$F = P(1+r)^T$
T years; n compoundings/year	$F = P(1+r/n)^{nT}$
1 year; n compoundings/year	$F = P(1+r/n)^n$

What about a value of F when we have an infinite number of compoundings:

$$F = P \left[\lim_{n \to \infty} \left(1 + \frac{r}{n} \right)^n \right] \qquad (2.1)$$

Here 'lim' stands for limit, because since infinity (∞) is a direction rather than a number, it cannot be inserted into (2.1), and we need to resort to something special, like approaching infinity but never quite getting there. However the answer is simple enough: $F = Pe^r$, where e is the base of the natural logarithm system, and it can be found on virtually every pocket calculator. Numerically it is approximately 2.7183. Now suppose that we put $1,000 in a bank that paid 10% interest, and kept it there for a year, with it being compounded an infinite number of times (i.e. continuously). We then have $F = Pe^r = 1,000e^{0.10} = \$1,105.2$. This should be compared with the values of F above when we have an infinite number of compoundings per year, but over T years: $F = Pe^{rT}$.

As an example let us take T=2, r=10%, and P = 1,000. We then get $F = Pe^{rT} = 1,000e^{0.10 \times 2} = 1,221.4$. This should also be compared with the earlier results. Something of interest here is that it appears that e^{rT} might function as an approximation for $(1+r)^T$. Let us try this for a one year time horizon (T = 1), with two values of r: r = 5% (= 0.05) and r = 50% (= 0.5); and with P = 1,000. Proceeding, we calculate two values of F:

r = 5%	$F = 1000(1 + 0.05) = 1,050$	$F = 1000e^{0.05} = 1,051.3$
r = 50%	$F = 1000(1 + 0.50) = 1,500$	$F = 1000e^{0.50} = 1,648.7$

In the first case the difference in percent is (1051.3–1050)/1050 = 0.0012 = 0.12%. In the second case the difference is (1648–1500)/1500 = 0.0987 = 9.87%. This tells us something: $e^{rT} \approx (1+r)^T$, when rT is small. The following exercises are easy, but they should be done anyway.

2.1.1 Exercises

1. Taking P = 1,000, what is F at the end of 10 years, if r = 5%, and n = 2?
2. What is F in the above exercise if continuous discounting is used? How good is the approximation? Explain effective rate of interest with an example where n = 12!
3. Show that: $\log_{10}1 = 0$; $\log_{10}0.1 = -1$; $\log xy = \log x + \log y$; $\log x^2 = 2 \log x$.
4. What is the exact meaning/definition of logarithm? What is the difference between logarithm (log) to the base 10, and a log to the base 'e'? What is the log of A
5. when $A = B^x$. What is the log of A when $A = e^x$, where e is the base of the natural logarithm system? (NOTE: *Natural* log is called 'Ln')
6. What is the log of 16 to the base 2? ($\log_2 = $?) What is Ln 16? Take $r_e = e^r - 1$. Explain!

2.2 A Very Small Oil Well

Suppose that Ms Bibi Sally sees a puddle of oil in her back yard one day, and comes to the conclusion that some methodical drilling is in order. She goes to the bank and requests a loan of 1,000 dollars to purchase a small drilling rig. She is offered a loan at the 'market rate of interest', which for this example is 10%. (We will leave for later the manner in which the loan will be repaid.) Her plan is simple. She will buy a rig, and hire her friend Bill to operate it for 2 years. Bill once taught economics at the Stockholm College of Economic Knowledge, and according to the boss of that noble institution he is mostly a charlatan; but even so Bibi agrees to pay him 300 dollars a year, especially after he informs her that she has a good thing, and can expect to extract 33 barrels of oil per year, which can be sold for 22 dollars/b. Bill likes 33 and 22, because they were his army serial number. (Furthermore, in the following discussion, never forget that 33 and 22 are *expected values!*)

But as usual, where economics is concerned, Bill is wrong. 33 barrels of oil selling for 22 dollars means 726 dollars/year, and it is collected at the end of each year. However this is Bibi's gross revenue. Her net revenue per year (i.e. her earnings) becomes 726–300 = $426, since Bill's salary must be taken into account. Let us call $426 the net profit, assuming no taxes, depreciation and depletion allowances, etc, for the time being. The present value of the two receipts of $426 is:

$$PV = \frac{V_1}{(1+r)} + \frac{V_2}{(1+r)^2} = \frac{426}{1.1} + \frac{426}{1.1^2} = 387 + 352 = \$739$$

The present value of Bibi's net receipts is only $739, and since the present value of the loan is $1,000, the investment is strictly a loser: Its NPV (= 739–1000) is less than zero. If she borrowed the money she would not be able to repay the debt at the end of the 2 years; or, for that matter, if she bought the rig with her own money ($1,000), her wealth would be diminished at the end of the second year – or in present value terms by $1,000–739 = \$261$.

But suppose that 33b/y could be extracted for 4 years, and sold for $22/b; and Bill not only will work for $300/y, but he will sweat and strain with the rig, night and day, to keep it in perfect working order during that period. The present value of the project's cash flows is now:

$$PV = \frac{426}{1.1} + \frac{426}{1.1^2} + \frac{426}{1.1^3} + \frac{426}{1.1^4} = \$1350.36$$

The PV is now more than the cost of the investment, and so NPV > 0: if she borrows the $1,000 she will be able to repay the loan at the end of the four years, or if she bought the rig with her own money, her wealth would be increased. In order to see what is going on here, it might be useful to examine the following diagram.

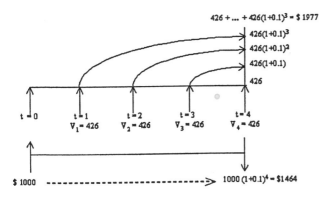

Figure 2.1

If Bibi puts her income flows (the V's = $426) into the bank as soon as she gets them, and receives 10% in interest, she will have more that enough to pay back her loan, which will require $1000(1 + 0.1)^4 = \$1,464$. To be exact she will have $1,977 (which the reader should verify), and so – measured at t_0, in terms of PVs; or measured at t_4 in terms of the F's – this operation paid a nice profit. Bill very definitely deserves a bonus for his work with the rig.

Two more things might be useful here. The profit in terms of the final value was $1,977-1,464 = \$513$ (why?). The profit in terms of present values was $1,350-1,000 = \$350$ (why?). Notice that $513 \approx 350(1+0.1)^4$. Let's also introduce a little sophisticated notation. For example, suppose that we have $G = X_1 + X_2 + X_3 + X_4$. This can be written:

$$G = \sum_{i=1}^{4} X_i = (X_1 + X_2 + X_3 + X_4) \qquad (2.2)$$

In this expression Σ (i.e. 'sigma') indicates a summation, and we say that the limits 'i' run from one to four. Now suppose that we have $Y = I + cI + c^2I + c^3I$. We can write this as:

$$Y = I + cI + c^2I + c^3I = I(1+c+c^2+c^3) = \sum_{i=0}^{3} c^i I = I \sum_{i=0}^{3} c^i \qquad (2.3)$$

Here we can recall that any number raised to zero is unity, and thus $c^0 = 1$. Next, let us introduce a summation sign into the example that we were working on earlier.

$$PV = \frac{V_1}{(1+r)} + \frac{V_2}{(1+r)^2} + \frac{V_3}{(1+r)^3} + \frac{V_4}{(1+r)^4} = \sum_{i=1}^{4} \frac{V_i}{(1+r)^i} = \sum_{i=1}^{4} \frac{426}{(1+0.1)^i}$$

Finally, suppose that Bibi and Bill come to the conclusion that the oil they are pumping will last forever, the price will never change, the amount will never change, and they will be alive forever to work the rig – which will last forever – and reap the profit. The present value of the operation now is:

$$PV = \sum_{i=1}^{\infty} \frac{V_i}{(1+r)^i} = \frac{V_1}{(1+r)} + \frac{V_2}{(1+r)^2} + \dots\dots + \frac{V_x}{(1+r)^x} + \dots\dots \qquad (2.4)$$

In other words, (2.4) is an infinite series. Question: what is the PV of this series, assuming that $V_1 = V_2 = ..= V_x = .. = V$? To begin, let us call $1/(1+r) = \theta$. Then we have:

$$PV = \sum_{i=1}^{\infty} \theta^i V_i = \theta V_1 + \theta^2 V_2 + \theta^3 V_3 + \dots\dots = \theta V + \theta^2 V + \theta^3 V + \dots \qquad (2.5)$$

Next we perform a simple operation on (2.5), noting that PV is written as (PV) when necessary in order to make sure that it is distinguished from the V's. Thus, starting with PV = θV + θ^2V + θ^3V + ..., where the right hand side of the last expression contains an infinite number of terms, we multiply both sides by θ, and obtain θ(PV) = θ^2V + θ^3V + θ^4V + ... Subtracting the last of these expressions from the first we get PV – θ(PV) = θV, or PV(1 – θ) = θV, which can be written (PV) = θV/(1–θ). Finally, making θ = 1/(1+r) once again, we obtain the very simple result that is shown in equation (2.6).

$$PV = \frac{\dfrac{1}{(1+r)}}{1-\dfrac{1}{1+r}}V = \frac{V}{r} \tag{2.6}$$

If we put the familiar figures into (2.6) we get PV = V/r = 426/0.1 = $4260. Everything considered, $4,260 does not seem like a great deal, but if you stop and think about it, $426 received in 50 years, and discounted at 10% is not very much either, as you can easily verify. Of course, 50 years does not amount to much in terms of historical time, and some economists have argued that governments should keep this in mind: to be specific, when they are considering future benefits, they should use very low discount rates, regardless of what discount rates are being used in the private sector. Thus, certain projects that might be rejected by the private sector when present values are considered (such as health care), might be found acceptable when scrutinized through the medium of extremely low discount rates. In fact, the late Frank Ramsey of Cambridge University was prone to argue that since countries, in theory, had infinite lives, governments should use a discount rate of zero.

Notice something else! The discounting above took place using interest rates: The bank offered Bibi a loan at a certain rate of interest – perhaps the market 'loan rate' for small borrowers (which might be double the 'deposit rate', or more). But she is not compelled to discount her future cash flows at this rate of interest, especially when these flows are uncertain. This aspect of her behavior is not a market related matter. She might have chosen to discount these flows at a much higher discount rate. She might have said to Bill: "Discount rates are subjective. You say that I will get $426 every year, but I'm not certain. I'll only go through with this project if, using a discount rate of 15%, I get a NPV > 0. You see, if I were *certain* that I would get

$426 every year for the next four years, there would be no problem; but with a discount rate of 10%, an *uncertain* $426/year for 4 years does not give me as much satisfaction as, e.g. my $1,000 invested in a week's vacation in Paris."

She might also mention that if Bill has studied capital budgeting, then he should be aware that the correct discount rate for physical investments is the *opportunity cost* of capital for a particular project. In plain language, this is the expected rate of return that could be earned from an investment having a comparable risk (i.e. is in the same risk class). The actual calculation here usually reduces to a weighted average of the expected return on a company's shares and the interest rate it pays for debt – assuming that the investment in question is typical for the firm as a whole. (However, it is not certain whether these observations apply to the not very imposing project under discussion in this section.)

In addition it should be noted that a low discount rate gives more weight to cash flows that are earned in the distant future, while a high discount rate gives distant earnings much less weight, and can make a manager appear myoptic in his or her evaluation of potential investment projects. The discussion in the last part of Chapter 9, however, will suggest that a resort to what appears to be unreasonably high discount rates may be the result of a manager's desire to delay the initiation of an investment project – i.e. to keep option's open – until more information about costs and benefits can be accumulated.

And finally, as alluded to at the beginning of the above discussion, most of the revenues and costs in the real world are *expected* revenues and costs. If we take the example that we have been working with, unless the expected oil price is hedged with e.g. a long term contract, it is not really certain. There may be a different quantity of oil on her property than Bibi thinks; and Bill may change his mind about working for $300/year. Thus, calculating and using PVs is more of an art than a science, and this is true even if such things as real option theory is brought into the picture.

2.2.1 Exercises

1. In the four period operation discussed in the above section, suppose that Bibi uses a 15 percent discount rate. Is that operation still attractive to her?

2. If the oil well above could be exploited forever, we would still only get a PV of $4,260. Forever is a long, long time. Why don't we have PV in the neighborhood of infinity? When would we have a PV of 'infinity'?

Strictly speaking, is it correct to use the expression 'PV is equal to infinity'?

3. What are the factors that would make Ms Sally decide *not* to exploit the oil deposit in her back yard, even if Bill assured her that it could be pumped forever, and she believed him?

4. Write out 4 terms in $Y = \Sigma\theta_i X^i$, making $i = 3$ and $i = 8$ the limits of the summation.

5. Bibi borrowed money for 10%, and her project was a success. What can you immediately say about the 'yield' or 'rate of return' on this project? In the four period example given in the section above, can you say what is the approximate rate of return? Use one of the numerical examples in the above section to make a detailed calculation of NPV, using the definition given in the first paragraph of the section!

2.3 Annuities

The obvious next step is to calculate the numerical PV for a finite – as compared to an infinite – monetary stream that displays a certain value of V in each period. This elementary undertaking should be attempted by the reader, who has only to reproduce the manipulations leading up to equation (2.6), although with a finite payments stream. For instance, use two periods, with payments being made at the end of the periods. In case you need some help, a similar exercise will take place directly below in this section.

In the previous section we examined Ms Sally's profits at time t_0 by comparing the PV of her investment with the cost of that investment: the profit in present value terms was $350. We did the same thing at time t_4, by comparing the final value of her accumulated profits, suitably invested, with the amount of money needed to repay her debt to the bank, including accumulated interest ($= 1000(1+0.1)^4$. Her profit measured at t_4 was $513.

Some terminology might be of interest here. The realization of a positive profit when monetary flows are discounted at the prevailing rate of interest, is tantamount to an increase in wealth. Remember the scenario in the previous section. Ms Sally sees the puddle of oil in her back yard, and comes to the conclusion that is originated in a larger deposit. But this is not certain. Given her expectations, had she bought a rig and was only able to drill two periods, earning (as profits) $426/period, her wealth would have decreased, because she would not have earned enough to repay her loan. But as we saw, a four period operation would have earned her enough to pay all expenses, repay her loan, and to have something left over. Thus we are entitled to say that her wealth has increased. In fact, were geology an exact science, and a geologist confirmed the size of her deposit, then in a mainstream textbook world she might have celebrated her good fortune by selling the asset (i.e.

deposit) on the spot, and departing for Paris, or maybe Courchevel or Åre to do a bit of skiing.

We go to the next main business of this section. Suppose Ms Sally borrowed an amount of money PV at time t_0. *Ceteris Paribus*, at time T an amount $PV(1+r)^T$ would be due the bank. Now, suppose she repaid her debt by an amount A each year. Exactly what Mr. Bank Manager did with his money (i.e. A) is a well kept secret, but for the purpose of computing an annuity the assumption is that as every 'A' was received, it was invested by Mr. Manager at the prevailing rate of interest, which for *notional* purposes is taken as 'r'. (Notional mean the amount or value that applies to a specific transaction, regardless of other circumstances. For instance, you could be repaying the debt on your house or car at the notional rate of 8%, although the prevailing rate of interest for consumer loans is 6.5%.)

Let us continue by examining the four period case we treated in the previous section, using the same figures, and perhaps referring to Figure 2.1. An annuity is the amount paid every period so that if it was invested by the receiver at the notional rate of interest, it would be sufficient to repay the debt $PV(1+r)^T = 1000(1+0.1)^4$ at the end of the *amortization* period. (Amortizing a debt means repaying the principal, which in this case is $1,000, with the amortization period being 4 years.) The algebra associated with this particular arrangement is:

$$PV(1+r)^4 = A + A(1+r) + A(1+r)^2 + A(1+r)^3 \qquad (2.7)$$

In this expression we can observe that the first term, A, refers to the payment at the t_4. If we multiply both sides of (2.7) by $(1+r)$ we get:

$$(1+r)[PV(1+r)^4] = A(1+r) + A(1+r)^2 + A(1+r)^3 + A(1+r)^4 \qquad (2.8)$$

Subtracting (2.8) from (2.7) we obtain:

$$(1+r)^4 PV[1 - (1+r)] = A - A(1+r)^4 \qquad (2.9)$$

Simple manipulation of (2.9) will result in:

$$A = \frac{r(1+r)^4}{(1+r)^4 - 1} PV \qquad (2.10)$$

We can now use figures from the previous section to obtain A = $[0.1(1+0.1)^4]1000/[(1+0.1)^4-1]$, and this is equal to $315.4. In other words, Ms Sally can repay her $1,000 debt to the bank by paying them $315/year (at the end of every year for 4 years), instead of $1,464 at the end of 4 years.

These annual payments could also be divided up into a payment on the principal, which originally was $1,000, and a payment of the interest. This matter will be clarified below with a simple example. If we recall that by selling 33 Bbl of oil per year for $22/Bbl she earned $426 after expenses, then her profit per year (before taxes) is 426 – 315 = $111. In case the reader has not guessed, this positive profit alone, calculated at a 10% rate of interest, indicates that the yield (or return) on her investment is more than 10%.

Something else that needs to be made clear is that if Ms Sally had bought the drilling equipment with her own money instead of borrowing it, we would have employed exactly the same type of calculation. This is because 'r' is an *opportunity cost*, in that whenever an investment is being considered in which the outcome is uncertain, an alternative is a 'safe' financial asset such as a bank deposit or a government bond. (Of course, the yield from a similar project might be regarded by a capital budgeting specialist as a better estimate of the opportunity cost, although for a small saver or investor, this is not particularly relevant.) Ignoring inflation, by investing in an oil well instead of a bank deposit, Bibi gives up a certain $1,464 at the end of 4 years; but as things happily turned out, her $1,000 grew to $1,977, assuming that she banked her income from the investment at the end of every year (as suggested by Figure 2.1).

What we can also say here is that her wealth increased: it increased by 1,977 – 1,464 = $513 in current value terms, or $350 in present value terms. (Why?) This is what successful investments are all about; and a more advanced argument in which wealth was reckoned in 'utility' units would have come to the conclusion that her wealth would have increased even if she had consumed her profit as soon as it was made.

Expression (2.10) is the value of a 4-period annuity. Suppose instead that we have T periods. As can be easily verified, our expression would become:

$$A = \frac{r(1+r)^T}{(1+r)^T - 1} PV \qquad (2.11)$$

This is one of the most valuable expressions in economics. Let us use it again to get an insight into the difference between the amortization and interest payments on an asset. For instance, suppose we buy a house for $1,000, with r = 10%. Assuming that we amortize our debt in 2 years, and calling this debt PV, we have for our annual payments 'A' (made at the end of the year) A = $[0.1(1+0.1)^2]1000/[(1+0.1)^2 - 1]$ = $526, and these are paid at the end of the year for two years. The following tableau applies to this example.

Year	Beginning Balance	A	Interest	Capital	End-of-year-balance
1	1000	576	100	476	524(=1000-476)
2	524	576	52.4	524	0

The mechanics of this calculation are as follows. Each year a payment of $576 is made. The capital payment (i.e. amortization) for the second year is $576 - 52.4 = \$523.6$ (≈ 524), since with a balance of 524 going into the second year, the interest payment (for that year) has fallen from 100 to 52.4. (Some 'rounding off' has taken place in this example). Finally, pay special attention to the sum of values in the 'capital' column. They sum to the principal (= $1,000).

In reflecting on our oil drilling example, we should discern that an investment is profitable if the discounted value (PV) of its profit flows are greater than the cost of the investment, which from now on we shall call 'I'. (This is the same thing as saying that NPV > 0.) With oil flowing 4 years in our example, we see that PV > I (= $1,000), which we also interpreted as meaning that the investment yielded more that 10%. Thus, in the remainder of the book, some effort will be made to distinguish between PV and I, which means between the present value of profit flows and the *investment cost*, where the latter is *not* the same thing as the capital cost. In the very unsophisticated situations that we examined above, 'A' was the capital cost: i.e. the periodic (e.g. annual) payment for the capital asset (i.e. the drilling rig).

Before the reader does a few exercises, something that is often overlooked needs to be given a little attention. This is the difference between *real* values and *nominal* (i.e. monetary) values, and how this difference fits into our previous discussion. Assume that you put $100 dollars into a bank, and today this money will buy 100 cucumber sandwiches – which is the only item produced in the country where you living. (Obviously we have for the price of cucumber sandwiches $P_c = \$1/\text{sandwich}$.) If the nominal rate of interest is 10%, then at the end of the year you have $110, and if the price level has not changed you can buy 110 cucumber sandwiches. We can say here that the real rate of interest – which is a commodity rate – is the same as the nominal rate. Both are 10%!

But suppose that there is inflation, and the price of cucumber sandwiches increased by 5%. In these circumstances the $110 will only buy 110/1.05 = 104.76 (≈ 105) sandwiches. The money rate of interest is 10%, but with inflation the real rate is 5% (= (105–100)/100). Had the rate of price increase been 10%, then the real interest would have been zero. Formally – but not exactly – we have: real rate of interest = nominal rate *minus* inflation rate. This idea also carries over to such things as income and wages. If your income goes up by 8%, but the inflation rate is 8%, your real wage increase is zero.

That brings us to the crux of this discussion. If the money flows in our discounting exercise are in real terms, then the discount rate should also be in real terms. The reason is that the discount rate is the (subjective) opportunity cost of an investment, and if inflation is not included in the cash flows that resulted from the investment, it should also be excluded from the opportunity cost. Correspondingly, if cash flows are in nominal (i.e. money) terms, then the same should be true of discount rates. Most of the time cash flows are in nominal terms.

2.3.1 Exercises

1. Suppose that Ms Sally could operate her oil well for 5 periods under the same conditions given above in regard to income received, Bill's salary, the rate of interest, etc. What would the annual payment (A) to the bank be? By how much has Ms Sally's wealth increased due to this successful investment? Can you calculate the effective yield of this investment (in percent)?
2. There is a tabular example in this section about buying a house for $1,000, and paying for it in two years (with r = 10%). Make that 3 years, and rework the table!
3. Suppose that Ms Sally borrowed $1,000 to buy a rig that could pump oil forever, and she and Bill did exactly that. What would her annual payments be?
4. Suppose she could only pump oil for 2 periods, but at the end of that time the drilling rig could be sold for the same amount that she paid for it ($1,000). Would it make sense now to undertake this venture?
5. If you borrow $250,000 to buy a machine, and the amortization period is 10 years, while the loan is amortized at 6% every 4 months, what are the payments?

2.4 Some Comments on Capital Values

It hardly needs to be said that there is more to evaluating investments than the simple discounting techniques considered above. Uncertainty often has to be brought into the picture in a meaningful way, and such things as the irreversibility of investments should be considered. But our study of these matters should begin with a solid understanding of elementary discounting procedures.

Let us continue our previous discussion on a slightly higher key. We should have begun to understand that the price of many physical and financial assets is determined from expectations about the future returns and costs. In addition, readers should have begun to glimpse the 'time is money' aspect of our work. If Bibi buys a rig and sells it four years later for $1,000, it still has a cost. The present value of $1,000 after 4 years is $1000/(1+0.10)^4 = $683, using 10% as the interest rate on a deposit or a bond. The cost is thus $1000 - 683 = 317, which is the present value of the interest income she loses if a physical or financial investment is available with a return (or yield) of 10%/y.

But note the following: throughout most of the above discussion, for simplicity, the premise has been that loan rates and deposit rates are the same, although this is definitely not true in the real world. If the alternative to buying the rig is a 10% interest income (e.g. from a bond), or if she borrows money at 10%, then the above calculation applies. But if she takes money from a bank account paying 5% in order to buy the rig; or e.g. if she wins $1,000 playing lotto, and uses this money to buy the rig instead of putting it in the bank to collect interest at 5%, then the above calculation is incorrect. If we continue to assume that in 4 years she sells the rig for $1,000, then her loss is $[1000(1+0.05)^4 - 1000]/(1.05)^4 \approx 177.

Next we come to depreciation, which can mean several things. Among these are actual physical depreciation (or deterioration), where a machine wears out and/or its productivity falls – gradually or suddenly. Very often, however, the depreciation cost – or better, *allowance* – refers to a government rule or directive for determining, for tax purposes, the notional amount by which the value of an asset falls every year, and which usually has nothing to do with physical deterioration. What this is all about is making it possible for the purchaser of a physical asset that is used in a business to replace that asset after its useful life is over. It can happen, of course, that provision is made for the replacement of the asset during a 10 year depreciation period, while the asset is still in productive use after 100 years, but that is another matter entirely. For the owner of the asset, the shorter the depreciation period the better, since a short depreciation period means a high depreciation allowance, which in turn reduces taxable income. In the examples that we have been working with above, the tax per period is simply t[pq – Bills's salary – D], where t is the tax rate, pq the gross revenue, and D is almost always referred to as the depreciation 'allowance', rather than the depreciation cost.

This might be the place for another numerical example. Suppose that Ms Sally uses $1,000 of her own money to buy a drilling rig that deteriorates at the rate of 10%/year. If this reflects on production, then *ceteris paribus* production during the second year will only be 33 x 0.9 = 29.7 barrels (which can be sold at $22/b). The value of the rig at the end of the first year will be taken as $900, and if it deteriorates another 10% during the second year, its value at the end of that year is $810. (Observe! These are assumptions for the purpose of this exercise. The value of the rig is, of course, its market value. If there is a shortage of rigs due to a drilling boom, then even though this rig has physically deteriorated, it might sell for several thousand dollars after 2 years. Note too that this kind of deterioration is *exponential*. Depreciation allowances on the other hand, are usually of the *straight line* variety.) Let us put these figures in a present value exercise of the type with which we began this chapter.

$$PV = \frac{33x22 - 300}{(1.1)} + \frac{29.7x22 - 300}{(1.1)^2} + \frac{810}{(1.1)^2} = \$1347$$

In this 2-year scheme we have PV > I (i.e. $1,347 > $1,000). The investment is profitable assuming that we really do get the number of barrels/year predicted, they sell at the forecast price, Bill continues to work with a high efficiency (and for the same salary of $300/year), and the market value of the drilling rig does not fall too much lower than the expected scrap value that is employed in the calculation ($810).

Taxes and depreciation for tax purposes can also be put into this example. Normally Bibi would have to pay a tax on any profits she might make. In the example above, let us assume that the tax rate is 20%, and so in the first year she would have to pay 20% of $426. But if there is a depreciation allowance, then before she calculated her tax she would be permitted to deduct this amount from the $426 before she calculated the tax. For instance, if the tax authorities specify a 5 year depreciation, and a straight line depreciation scheme, then Bibi would enjoy a deduction for tax purposes of $200/year. Her taxable income for the first year would now be $226 instead of $426.

In the second year we have a slight problem, because tax rules are not the same in every country; but it might be so that her gross income in that year was the amount she received from the sale of oil (pq) *plus* the entire resale value of the drilling rig. To get her taxable income, Bill's salary and the depreciation allowance for that period would be deduced from that amount. Her profit/year would then be (pq − Bill's salary − taxes + resale value of equipment).

Before moving to a new topic, it needs to be pointed out that impressive advances are now being made in oil and gas exploration technology, but even so it appears that these improvements have had less impact on the net cost of drilling than the interplay of supply and demand in the rig market. When oil prices slumped in 1998, some rigs that had rented for $160,000/day were on offer for $30,000/day. One of the lessons here seemed to be that large drilling cost efficiencies might be realized by exploring and/or developing when rig rental rates were low, instead of following the usual practice of waiting for high oil prices to initiate these activities.

2.4.1 Exercises

1. Use some of the figures above to calculate a PV of Bibi's investment! In the 'time is money' discussion above, what conclusion should be drawn if the interest rate is zero?
2. Comment on the advantages (or disadvantages) of using straight line depreciation for tax purposes!
3. Distinguish between deterioration and depreciation!

2.5 Depletion

One of the things that a good book in energy economics must do is to make sure that its readers obtain a comprehensive vocabulary, so that he or she will not appear to be out of place when in the company of energy professionals.

Depreciation is a term well known in economics, but not depletion; and in the great world of Texas and Oklahoma (USA) oil, depletion seems to have been as important as taxes, subsidies, and depreciation – especially for politicians who wished to continue in that career. Depletion provides for the recovery of money invested in a wasting asset in the same manner that depreciation allowances are intended to do this for a physical asset. The correct expression is, of course 'depletion allowance', but in the vernacular 'depletion' is employed. The gist of all this is that in addition to other deductions, an annual depletion allowance is subtracted from gross revenues before computing taxable income. One common method is called 'depletion by cost' (DE_t) for the amount q_t in the year 't'. Here we have:

$$DE_t = \left[\frac{Cost \quad of \quad property}{Total \quad units \quad in \quad property} \right] q_t$$

Another approach to obtaining the annual depletion for tax purposes for oil, gas, and non-fuel mineral properties is to use a percentage of gross income. In the US, for example, this is 22% for gas and oil, and 10% for coal. A further stipulation is that the amount allowed for depletion is less than 50% of the taxable income of the property before depletion.

If producers have a choice, they understandably choose the arrangement that will result in them paying as little tax as possible. In order to explain the issue, let us employ the kind of exercise that was discussed earlier in the chapter, although we will make certain modifications for the purpose of keeping everything as simple as possible.

The assumption now is that there are only 66 units (i.e. barrels) of oil in the property, and that the cost of the property is only $500. As before, production is 33b/y, which can be sold for $22/b. Equipment to obtain this oil costs $500, and Ms Sally envisages it having an appreciable scrap value. The tax people, of course, are not concerned with projections of the scrap value, and according to their regulations, the equipment can be depreciated on a straight line basis over 5 years. Ms Sally's friend Bill Lather will operate this equipment as before, and he will be paid $300/y in both years. The tax rate is 20%.

The annual revenue in both years is $pq = 22 \times 33 = \$726$. The depreciation allowance is $100, and as for depletion we have for the first year: $DE_1 = (500/66) \times 33 = \$250$. We thus have for costs and benefits (i.e. revenue) for that year:

	Costs	*Revenues*
Gross income		+726
Variable costs	−300	
Depreciation	−100	
Depletion	−250	
[ΣCosts, ΣRevenues]	−650	+726
Taxable income		$76
Taxes:	0.2 x 76 = $15.2	

Now let us investigate the 22% depletion rule:

	Costs	Revenues
Gross		+726
Variable costs	−300	
Depreciation	−100	
[ΣCosts, ΣRevenues]	−400	+726
Taxable income before depletion		$326
Depletion allowance:	0.22 x 726 = $159	

This is 22% of gross income, as specified above, and since it is less than 50% of taxable income before depletion, it could be used in a calculation of taxable income – although it would be irrational in the present example, since depletion by cost provides a more favorable tax situation.

It might be useful to see what the addition of a depletion allowance does for the attractiveness of this 2-year project. For the (expected) PV we have:

$$PV = \frac{(726 - 300 - 15.2)}{(1+r)} + \frac{(726 - 300 - 15.2)}{(1+r)^2} + \frac{Equipment's \quad terminal \quad value}{(1+r)^2}$$

The assumption here is that the land (and its value) drops out of consideration after the project has been terminated, while the drilling rig has a terminal value. For tax purposes the rig has lost 2/5 of its value, but as far as the market is concerned its value is $350. It could happen that a tax has to be paid on the sale of this equipment, but the assumption at this time is that the tax is zero. With $r = 10\%$, the PV of this investment is $1,001.25, which the reader should check.

In other words, if $I = \$1,000$ (= cost of land + cost of equipment), then this investment is barely profitable: $PV = \$1,000.25 > \$1,000 = I$. This is the kind of project that needs a very careful scrutiny by an expert before it is approved. Employing the NPV approach, it does not look promising.

But notice, real option theory suggests that NPV analysis can fail to capture all the sources of value associated with various types of investment opportunities – e.g. when irreversibility and certain types of uncertainty are present. Carrying out the above project might lead to information or knowledge of the sort that can expand the value of asset under consideration, or for that matter similar assets. It might also increase managerial flexibility, which in theory has a calculable value. For instance, since oil prices are highly variable, then if this project were undertaken and the oil price suddenly escalated, it might be possible to sign a long term contract for all or a part of future output, at or near the higher price. Readers should remember this example while they are reading Chapter 6, since making the above investment then becomes analogous in some ways to buying a *call option*. "In some ways", because a more typical example would involve purchasing reserves and leaving them undeveloped until the price escalated. (The cost of this purchase would be called a 'premium'.) An investment would then be made in developing the reserves if oil price movements were favorable, with, perhaps, a long term contract involved in the manner alluded to above. (The cost of development would be termed the 'exercise price', and thus the actual development is tantamount to exercising a call option.) This kind of terminology and thinking will become second nature by the time you finish Chapter 6, although real option theory is taken up in Chapter 9.

We can wind up this chapter with a few remarks that should be useful to many readers. In the beginning of this chapter, we looked at expressions such as $p_1 = p_0(1+r)$: if you put p_0 dollars in a bank, and the money rate of interest was r, then at the end of the year you would have p_1 dollars. Suppose that instead of money you had a bar of precious metal, with a price of p_0 at the present time, but which will – with certainty – have a price of p_1 in a year. We can then define $(p_1/p_0)-1$ as the *own rate of interest* of this commodity over the interval $(0,1)$. If we assume that $r > [(p_1/p_0)-1]$, then if you have the commodity you can sell it for p_0, invest this money in a bond or bank account carrying an interest rate of r, and after one period you can repurchase the commodity, and have something left over. But note: you must be sure of p_1!

This kind of operation is an (perhaps too simple) example of arbitrage, and it implies a riskless profit. Presumably, it would be carried out until $r = [(p_1/p_0)-1]$, which is called the *no-arbitrage* condition. (And why would this 'equilibrium' be reached? The answer is that with the possibility of doing the arbitrage mentioned above, all producers of the commodity would produce and sell it as fast as they could, just as all inventory holders would reduce their stocks as soon as they could. These actions could be expected to depress p_0, and *ceteris paribus* this would restore the no-arbitrage condition.)

Note also that if we rewrite the no-arbitrage condition as $r = (p_1-p_0)/p_0 = \Delta p/p$, we get Harold Hotelling's so-called 'pathbreaking' relationship (1931). Hotelling's work will be referred to later, but readers should already be able to see that it is not particularly enlightening. On this point, Roncaglia (1994) has some very cogent observations.

2.5.1 Exercises

1. Construct a problem of the type used in this chapter where we have depreciation, taxes, and the 22% depletion rule applies. Use a two period time horizon if possible, but if it is not possible use $T > 2$. Let the equipment used to obtain the oil or gas have a terminal (or scrap) value!
2. Suppose, in the discussion immediately above, we had the own rate of interest of the commodity larger than the money rate of interest. Describe the arbitrage that would take place!
3. The example of arbitrage given above is not particularly suitable for an introductory textbook. Provide a simpler one, making sure that you point out the no-arbitrage condition, and how it would be reached!

2.6 Conclusions

This chapter is fairly simple and straightforward, but one thing should be kept in mind. Calculations based on *expected* future revenues and costs can often paint a false picture of the profitability of investments. One of the ways that energy companies attempt to reduce some of this uncertainty is through long term contracts, which to a certain extent guarantees their revenue streams. But even so, in the energy field (and especially with oil and gas), executives can get some bad surprises where costs are concerned.

Real (or management or strategic) options theory involves trying to bring some sophistication to the NPV techniques emphasized in this chapter. A brief discussion of real option theory is given in Chapter 9, and the opening chapters of the book by Dixit and Pindyck (1993) should be useful to some readers of this book. Personally, however, I would be careful not to invest too much time in real options theory – unless, of course, you have nothing better to do with your time: differential equations are very useful for the study of such things as thermodynamics, or the trajectories of recoilless rifles and mortars, but they have very little, if anything, to offer economic theory. For more observations on discounting and/or uncertainty I can recommend Chapter 15 of Pindyck and Rubinfeld (1992), and Chapters 10 and 11 of Varian (1993). And don't forget, the last part of Section 8.2 provides more information on considering the choice of a discount rate for investment projects. Some readers will undoubtedly find it useful to turn to that discussion immediately.

In perusing the examples in this chapter, some sticklers for precision may want to know what happened to the opportunity cost of Ms Sally's and Bill Lather's time, where by opportunity cost they mean the income or utility they sacrificed in order to manage her oil venture. By way of avoiding this topic, which down-to-earth readers might find too esoteric for their tastes, I assume that Bill's career at the Stockholm College of Economic Knowledge was abruptly terminated – though not at his wishes. Upon taking notice of his fall from academic grace, Ms Sally immediately ended her business partnership with that gentleman, citing the large number of psychiatric disorders that tend to be characterized by unsuitable social and professional attachments. Besides, she was the lucky beneficiary of a trust fund established for her by a rich uncle, who wanted her to have the leisure to improve her knowledge of energy economics. According to that gentleman, the Nobel Prize chemist Frederick Soddy always insisted that "the flow of energy should be the primary concern of economics", and in the 21st century this might well be the case.

Chapter #3

The World Oil Market

Although some economists still enjoy insisting that oil is inexhaustible, geological authorities are increasingly expressing the belief that this resource is painfully finite in an economic sense, because in some parts of the world it might become too costly to produce in the amounts supplied earlier.

For example, within many traditional producing regions, a (large) unit of oil now tends to be more difficult to find and costly to extract than previous units. Put another way, the larger the oil field the more likely it is to be discovered early; and since large oil fields, often called 'elephants', contain a disproportionate amount of global oil, discovery rates have shown a tendency to decline, and the content of fields to decrease, as time passes – *ceteris paribus*. In the last decade this decline has become highly visible, and as Cleveland and Kaufman (1993) have shown, drilling costs are rapidly increasing in many parts of the world.

But why ceteris paribus? The answer is that the world oil scene is being divided into two arenas: the Middle East and the rest of the world. The reference case projection of the US Energy Information Administration (EIA) now features the Persian Gulf managing two-thirds (or more) of the world's exportable oil by the year 2005, with the United States importing 60% of its consumption for a cost of 100 billion dollars per year. More important, although the cost of discovering and exploiting oil fields outside the Middle East may be escalating at an unpropitious rate, there should still be plenty of low-cost oil in the Middle East, although its owners will probably not make it available at bargain basement prices.

In 1995 Colombia topped the reserve finding league, with 755 Mb added (as reported by the consulting firm Petroconsultants). The US came second, with 666 Mb, and Algeria third (504 Mb). But even so, it was the first year in modern times that new reserves failed to exceed 10 Gb on a global basis;

and 1995 was apparently the tenth year in a row that additions to world oil reserves declined. The following year the growth in oil consumption leaped to its highest rate in a decade, but the pressure that this put on the oil price was alleviated to some extent by increased production from Norway and Canada. The problem is that by 2005, or earlier, these two countries – among others – may not be able to raise their production to such an extent that they will be able to fill large potential gaps between global oil demand and 'intended' supply. ('Intended' because OPEC attempts to set a production ceiling.) On those dramatic occasions, there will be an irresistable call on OPEC for more oil. Their response could be of crucial importance for the entire world economy.

A few oil demand forecasts are given below. No price forecasts are given because, like the present author, the organizations mentioned have been greatly embarassed by the actual oil price deviating sharply from the one predicted. But notice the situation in the year 2010. By that time the North Sea will be a minor oil province; US imports will be huge; oil consumption in Asia could still be expanding by a very large amount; Australia and Canada's oil sectors will be long past their best days, and according to Leo Drollas – chief economist of the Centre for Global Energy Studies (London) – so will those of Qatar, Libya, Algeria, Gabon, Nigeria, and Indonesia. This might also be the place to note that the forecast of the International Energy Agency (IEA) of the OECD, presented at the 1999 international meeting of the International Association for Energy Economics (IAEE) stated that non-OPEC production will peak before 2010, and global production will peak before 2020. The units in Table 3.1 are millions of barrels/day (Mb/d).

Table 3.1

Forecaster	Base Year (1994)	2000	2010
IEA	68.3	76.5	93.5
OPEC	65.7	72.3	—
US DEPARTMENT OF ENERGY POLL OF FORECASTERS	68.6	76.4	86.3
US DEPARTMENT OF ENERGY	68.6	78.6	88.7
Average	67.8	75.9	87.2

Source: IEPE (Grenoble)
Note: Estimated Global Output for 2020 by the IEA: 100 Mb/d.

If private automobile ownership in Central and Eastern Europe, and East Asia, continues to advance toward lower Western European levels, then by 2010 global oil demand could increase by a much larger amount than given above. That possibility causes me to recall 1973-74, when Occidental Oil's colorful Armand Hammer openly insisted that by the end of the 20th century, oil would be selling for $100/b – a price which, if realized in 2010, or even 2020, could be a very bad shock for a large part of the oil importing world. Needless to say, given the information that we possess at the present time, this kind of price seems very unlikely – but no more unlikely than the oil price collapse that certain Nobel Prize winners in economics once predicted.

At the same time, when we begin to talk about 'alternative' motor fuels, it needs to be remembered that an input for some of these alternatives, natural gas, might not be as plentiful as we hope it will be in another 20-25 years, due to its accelerating use in the generation of electricity. (In addition, at the beginning of the 21st century, gas-to-oil technologies are supposed to be at the 'take-off' stage.) By way of contrast, it might be possible to introduce some version of an 'electric' or 'hydrogen based' automobile on a fairly large scale in a relatively short time, but unfortunately 'fairly large' is not good enough.

Teitelbaum (1992) quotes the 'universal investor' Marvin Davis as saying "You don't have to be a cockeyed genius to see this coming" He might have added, 'it doesn't make any difference when it comes – any time will be too soon'. Oil is the most important commodity in the world, and will remain so for the indefinite future. I therefore submit that even a wrong forecast, if it is on the high side, should be treated with the greatest respect. That, to me, is what risk aversion is all about.

3.1 Some background for Studying the Run-up to the Oil Market Endgame

The 1996 meeting of the World Energy Council (WEC) highlighted a situation of which many of us have long been aware: the demand for energy, and especially oil, will be increasingly tied to the size of the global population, while the supply of oil will eventually turn on the strategy adopted by Middle Eastern producers. This reminder can also be found in an extremely perceptive short (and non-technical) article by Howell, Bird, and Gautier (1993). Their position is that there is an embarassingly limited amount of conventional oil in the earth's crust, and it is not particularly wise to suppose that this resource can be greatly augmented by marginal increases in the price of oil. Put another way, increases in the price of oil could promote, possibly by a large amount, the search for oil, but unfortunately the searchers cannot find something that does not exist. Technological progress must somehow come to the rescue, and a very large number of economists say that it will. What they do not say, however, is when and how – nor should they be expected to be in possession of this information. Unfortunately, there is no such thing as a credible *microeconomic* theory of technological progress.

This might also be a good time to recall the way that the great philosopher Karl Popper viewed future knowledge. He said that it impossible to predict future developments in the state of knowledge, since such predictions turn on the ability to describe various pertinent features of such knowledge – which is only possible if one possesses that knowledge already. We know from past experience, for instance, that science is capable of presenting us with miracles, and it makes sense to expect that it will continue to do so throughout the 21st century; but just as important, we cannot say a great deal about the form of these marvels, nor whether they will appear when they are most needed,

Many economists enjoy believing that economic and, to a lesser extent, political factors are paramount in the determination of the oil supply, and thereby its price. (This includes high-profile energy economists such as Morris Adelman and Peter Odell). These items are indeed important, but one of the things that will be strongly emphasized in this chapter is the overriding importance of geology, which is a reality that for the most part has been comfortably understressed in the economics literature. The truth of this contention will undoubtedly be made clear in a decade or two to everyone who is interested in the great world of oil, although it would probably be better for all concerned if it could be assimilated as soon as possible. It needs to be appreciated, however, that once non-OPEC oil production has peaked, political factors will indeed assume an overwhelming importance.

Some of the topics that will be taken up below have started to receive a great deal of attention in the business and financial press, while the recent papers of e.g. Criqui (1991) and Professor Paul Stevens (1995) indicate that academia would do well to consider moving energy economics to a more prominent position in the scheme of things. Quite naturally, there is no shortage of economists who will continue to insist that the price of oil will essentially mimic the past, with that price bouncing along at a level that is close to the one experienced in the middle of the 1990s, but with occasional shocks and 'bubbles', where a bubble is the departure of an economic variable, such as a price, from the value that is ostensibly justified by (supply-demand) fundamentals.

However, as already expressed, the opinion here is that the production and reserves picture that we have witnessed in the US over the past decade will eventually be duplicated by most oil producing countries outside OPEC, as well as by a few members of that organization, and when this takes place the oil consuming world will be facing an entirely new oil supply picture. Among the many serious observers who now share this opinion are Daniel Yergin, and the former US Energy Secretaries Donald Hodel and James Schlesinger.

Here are a few other things for readers to keep in mind. There were almost 50 million more cars on US roads in 1995 than there were in 1975, and 750 million more miles were being driven annually. Vehicle and motorcycle ownership are exploding in some parts of the developing world (e.g. China and India). In addition, such things as accelerating urbanization will boost per-capita energy consumption. (In 2000, 43% of the world population was 'urbanized', while in 2015, from a much larger population, estimates are that this figure will increase to 55%.) This might also be a good place to mention that between 1970 and 1995, the consumption of commercial energy in developing countries more than tripled, and almost reached 30% of the world total. According to the World Bank, energy use in the developing countries will dominate the global energy panorama in the first decades of the 21st century.

Many readers of this book probably already know that a large number of surprises have taken place in international oil markets, and this will probably be true in the future; but it is clear that one of the most pleasant surprises for oil consumers has been the performance of North Sea producers – both in the UK and the Norwegian sectors.

Sometime early in the 21st century, perhaps very early, North Sea production should begin to slide, and if the oil price is not moving up, then offshore output could fall rapidly. (A rising oil price will promote more investment in marginal fields). Discovery in the Norwegian North Sea peaked in 1981, and it seems unlikely that a peaking of production can be delayed much after 2000. In the UK the most modern 'stand-alone' production platform ever constructed has been floated away from the Barmac fabrication yard (Scotland), beginning a 22 year stay in the central Graben area of the North Sea. Its cost is $2.33 billion, and it will be exploiting an estimated 700Mb of oil. Already it is being said, however, that few new discoveries have been made of a size to justify the construction of new stand-alone platforms. In fact, what is happening in the North Sea at the beginning of the 21st century should make it obvious that everything on the oil scene ultimately depends on the presence of the commodity in question. Technology has improved to a point where it is profitable to exploit even very low quality fields, but even these fields must contain a minimum amount of oil, which may not be the case in the North Sea much longer.

Although the implication in what follows is that oil will be recognized as a scarce commodity by the year 2015 or 2020, the not-so-perceptible-truth is that it is scarce now, but it is still impolitic to take this position. Instead, the actual degree of scarcity must be disguised by what Paul Tempest (1996) has called a "paradox": the great majority of investment spending on oil is taking place in high cost, relatively oil poor regions of the industrial world, whose reserves are thus depleted much faster than elsewhere. As Tempest makes clear, a consequence of this behavior is that even sophisticated observers are "dazzled by bouyant production growth among non-OPEC producers". Going one step further, the kind of economics usually taught in e.g. Scandinavian classrooms suggests that given the information we now possess about oil supply and demand, a steady upward movement in the price of oil would not be out of line with conventional neo-classical economic thinking. It would restrain the growth of consumption, promote investment in oil producing facilities, rekindle an interest in conservation and substitution, and speed up the development of new technologies. At the same time it must be admitted that excessive increases in the oil price could have far reaching macroeconomic consequences of a negative sort, as alluded to in the first chapter. This has often been an important element in the proposals for a dialogue between the oil producing and the oil importing countries – a dialogue that was suggested by the former, but rejected by the latter.

The next section of this chapter will be concerned with the reserve-production ratio, and as simple as this concept is, it is of crucial importance. It is *not* of "crucial importance" because it tells the entire truth about the oil market, or even a portion of it, but because it is one of the few workable analytical approaches to that very special market. For that reason the discussion has been carried out with a minimum of algebra, since there is a powerful lesson to be learned: when the reserve-production (= R/Q) ratio of the main oil producing countries outside the Middle East begins to approach 10, or thereabouts, the age of inexpensive oil could be just about over, and this is true regardless of what we see or hear about the 'life' of oil reserves, which has been defined (incorrectly) in the popular press as total reserves divided by annual oil production (R/Q). This is so because the output of a typical oil field will begin to decline when about half of the reserves they contain have been extracted, and production cannot be sustained unless more oil is found in the field, and/or new technology makes a more intensive exploitation possible.

This is an extremely important point, and so is the following! In 1956 M. King Hubbert began to publish research which indicated that production in the lower 48 states (of the US) would peak sometime between 1965 and 1970, although there would still be an enormous amount of oil in the ground. (His empirical work centered around the use of a logistic equation of the form $Q_t = Q/(1+\alpha e^{-\beta t})$, where Q_t is cumulative discoveries or cumulative production (at time t), and $\bar{Q}$ signifies ultimately recoverable reserves. 't' is time measured with respect to some origin t_o, and α and β are parameters. A simple mathematical manipulation of the logistic equation after α and β have been estimated, will yield a Bell-like figure whose peak is easily distinguished.) The actual peak for 'lower-48' US production came in 1970, and since that time production in that region has been steadily falling.

Economists have almost uniformly ignored this apparent *tour-de-force* because Hubbert did not develop a formal theoretical framework that contained economic and/or political variables. Hubbert, in turn, claimed that these variables were reflected in the historical data he used; and in addition he challenged the economists to produce a model of their own which was as successful a forecasting tool as his construction – which, of course, they could not. Furthermore, Hubbert's work contained another piece of bad news that economists today make a special point of deprecating: even if technical change is capable of creating the equivalent of the kind of super-giant oil field that the best geologists employed by the richest oil companies can no longer find, it will not mean very much in terms of delaying the peak – given that oil production is as large as it is at the present time, and growing at its present trend rate.

If economists as a group were not interested in R/Q ratios, nor the work of Hubbert, then whose work were they interested in. Prior to the first oil price shock in 1973, the most important name in energy economics was Professor Morris Adelman of MIT, whose work on petroleum attracted international attention. After the shock Professor Adelman became more important than ever, but the economist whose research attracted the most attention (as measured in citations and extensions of his work in prestigious journals) was the late Harold Hotelling (1931). One of the claims of this book is that seldom in the history of academic economics has so much explanatory power been attributed to an approach that has so depressingly little to offer from a scientific point of view. (See also Roncaglia (1994) and the references cited there.)

Unfortunately, however, the Hotelling bandwagon kept rolling right along until the last years of the 20th century, and so we should not exclude the possibility of a precious few nuggets of truth in the work of Professor Hotelling and his disciples. For example, in a valuable article, Nordhaus (1973) has suggested that if the (net) price of oil persists in remaining below the Hotelling equilibrium path – where, as noted in Chapter 2, it should be increasing at a rate equal to the rate of interest (or $\Delta p/p = r$) – then at some point in time it will mushroom upwards. This statement is, to some extent, analogous to Laherrere's assertion (1995) that if efforts to delay the decline of world oil production "from around 2000" are sucessful, the decline will become a "cliff" (instead of a smooth fall). Such being the case, it might be tasteless to completely exclude Professor Hotelling's work from this book, but it will be relegated to the first section of Chapter 9. The best non-technical discussion of the Hotelling model can be found in Salant (1995), whom I prefer to think of as the most perceptive economist dealing with these matters – although I do not agree with his statement that the "framework" of Hotelling's model should be of importance to policymakers and analysts.

A very important oil topic will be treated in Chapter 6: oil futures, options, and swaps; and a non-technical preview of futures markets will be provided below, since it will facilitate the reading of the section on short-run pricing. Let me also suggest that if you should begin to have trouble with the algebra below, then you should immediately go to this elementary discussion of oil futures. But stick with the algebra as long as you can. It might come in useful some day.

3.2   The Reserve-Production Ratio

An important part of the story about the future of world oil can be determined by examining the history of oil in the United States, while keeping in mind the work of Hubbert described above. Amazingly enough, at a time before World War II when the supply of oil in the US was considered unlimited, Spencer Tracy, in a film whose name I have forgotten, told Clark Gable that it had taken millions of years to assemble the oil found in Texas and Oklahoma, and so it should not be recklessly squandered. At the present time the amount of oil remaining in those states is vaguely understood to be limited, but unfortunately there are no Spencer Tracys available to periodically inform the television audience of that sad fact.

This section introduces some elementary materials on the reserve-production (R/Q) ratio. This ratio is sometimes regarded as the basic numerical indicator of the production capacity of an oil deposit or oil field, or even a collection of fields. It is the ratio of remaining reserves to annual production, and a useful rule is that if the R/Q ratio falls to less than 10, then the deposit is being destroyed in much the same manner that sucking too hard on a straw destroys an ice cream soda, or driving a vehicle too fast causes excessive depreciation. (Some specialists say that when the R/Q ratio falls below 10 – or a figure near 10 – the field is being 'worked too hard'). It has happened that in the US, in the interests of conservation, laws have been passed prohibiting outputs that cause R/Q ratios to decline to less than 9 or 10. Readers should also notice the following: in theoretical economics profit maximizing producers are expected to behave in such a manner as to keep their assets from wearing out too soon, and thus the R/Q ratio deserves to be regarded as an economic as well as a geological parameter.

Before beginning, however, something must be made clear. To my way of thinking, the best – and perhaps the only – way to start thinking about the supply of oil is via the work of M. King Hubbert, as well as the wider significance of the R/Q ratio. But this is not all of the story: there are other factors that need consideration. The problem is that unlike our models in mainstream economic theory, these other factors did not take well to systematization, and so it has not been possible to build the comprehensive theoretical or empirical model that we need to describe in detail the functioning of this most important of all commodity markets, nor is such a model likely to appear in the near or distant future. This, I think, has been recognized for a long time; but since we would sincerely like to have a model, the Hotelling model was promoted (or, actually, overpromoted). Unfortunately, it also happens to be true that most of the persons actually working with oil – from the occupants of the executive suites in firms with revenues as large as the GNPs of many countries, right down to 'wildcatters' scouring the 'boondocks' for new strikes – would be unable to keep a straight face if confronted with the preaching of economists about such things as Hotelling's theory of depletion or the unlimited potentialities of derivatives markets.

Toward the end of the 1980s it appeared that the US was the only major oil producing area in the world where the R/Q ratio was under 10, and the opinion of many of us was that sooner or later the decline in the ratio would be checked by a precipitous drop in the production of oil. This happened in 1989 in the form of one of the largest production decreases in that country in modern times. (But remember from the previous section that production began to fall in the lower 48 of the US in 1970, when the aggregate R/Q ratio for these states was larger than 10.)

The question now is when will the same thing happen to that aggregate of countries outside the Gulf core of OPEC. The world-wide R/Q ratio seems to be about 41, depending on how reserves are defined and measured; but in the Gulf it is close to 85. During the 1980s OPEC increased its production by 20%, and its reserves by 75%. Reserve increases of this magnitude were not been noted in the 1990s, but apparently this is because prospecting in the Middle East has diminished.

Before we extend this discussion with some numbers, suppose we look at what appears to be an exception to the above rule. According to BP-Amoco's Review of World Energy for 1999, the R/Q ratio in the UK North Sea is 5.2. How can we explain this?

Probably the most important explanatory factor is the level of production that is necessary in order to justify the exceptionally large amount of exploration and development expenditures that have and will take place in that region. Without this production, it has been judged that existing assets –

which in this case means reserves *plus* production facilities (such as platforms, pipelines, etc) – would not be optimally exploited. When the problem is looked at in terms of short run profit maximization, and the present price and price expectations are taken into consideration, it could be argued that in some cases an 'overworking' of offshore deposits that were the object of very large investment expenditures makes economic sense to the firms and shareholders who are directly concerned. There are also some macroeconomic issues that are important here. These include the importance of North Sea employment, especially to Scotland; the importance of oil to the UK trade balance; and the importance to the UK government of tax revenues from North Sea oil.

Now for a simple numerical exercise. Assume that we have a field containing 225 units of oil, and we desire to lift 15 units/year. Since 15 is less than one-tenth of the reserves in the field, we can remove 15 units every year for 5 years (measured at the end of the year) without violating the above criterion. During this time the R/Q ratio(= θ) falls from 14 (at the end of the first year) to 10. (This can be denoted by the sequence 210/15, 195/15,.........,150/15). But after the fifth year, if we continue to remove 15 units/year, then we would be removing more than one-tenth of the deposit per year. (Note that by designating the R/Q ratio of 10 as the·critical R/Q ratio (=θ^*), we are saying that no more than 10% of the deposit should be removed in one year).

After the fifth year, θ^* determines production. In order for θ not to fall below 10, production in the sixth year cannot be larger than 13.64. This may not be completely obvious, so consider the situation during the 6th period (i.e. year). The R/Q ratio is constrained by a critical ratio of 10, and so (R_5–Q_6)/Q_6 = 10. With R_5 (reserves at the end of period five) equal to 150, we get $Q_6 = 13.6364 = 13.64$. This discussion can be easily generalized to:

$$\theta^* \leq \frac{R_t}{Q_t} = \frac{End \quad of \; period \quad reserves(t-1) - Q_t}{Q_t}$$

Replacing reserves by R we get:

$$Q_t \leq \frac{R_{t-1}}{(\theta^* + 1)} \tag{3.1}$$

In the example given here we can calculate the maximum production for the 'next' year, or Q_7. R_6 is $150 - 13.64 = 136.36$, and so with $\theta^* = 10$ we obtain $Q_7 = 136.36/11 = 12.4$ units.

Now let us turn to another example – or, more correctly, counterexample. This time let us consider a region, or for that matter two regions, or two fields having different sizes, containing at time t_0 reserves R_1 and R_2. Let us now assume that in the first field reserves are being depleted at a rate of Q_1 units per year, and after t' periods we arrive at the critical R/Q ratio θ_1^* for this field. That is, at $t = t'$, Q for this field (or deposit) begins to fall. Thus:

$$\frac{R_1 - t'Q_1}{Q_1} = \theta_1 = \theta_1^* \qquad (with) \qquad t > t' \Rightarrow Q_1 \downarrow \qquad (3.2)$$

Going to the second field, which contains R_2 of reserves at t_0, and with Q $= Q_2$ units/year, we want to look at the value of θ_2 at time t', assuming that Q_2 has not begun to fall: at time t' we will take $\theta_2 = \theta_2'$, and specify that θ_2' $> \theta_2^* (= \theta_1^*)$. By definition, the R/Q ratio of the field at t' is:

$$\frac{R_2 - t'Q_2}{Q_2} = \theta_2 = (\theta_2') \qquad (3.3)$$

We can now write as the aggregate R/Q ratio (= $\bar{\theta}$) for the two fields at time t', the following expression:

$$\bar{\theta} = \frac{(R_1 + R_2) - t'(Q_1 + Q_2)}{Q_1 + Q_2} = \frac{(R_1 - t'Q_1) + (R_2 - t'Q_2)}{Q_1 + Q_2} \qquad (3.4)$$

This relationship, along with (3.2) and (3.3), allows us to write:

$$\bar{\theta} = \frac{\theta_1^* Q_1}{Q_1 + Q_2} + \frac{\theta_2' Q_2}{Q_1 + Q_2} = \beta_1 \theta_1^* + \beta_2 \theta_2' \qquad (\beta_i = \frac{Q_i}{Q_1 + Q_2}) \qquad (3.5)$$

The betas (βs) obviously add to unity, and since a condition is that $\theta_2' > \theta_1^*$, then we must have when production begins to fall:

$$\theta_1^* < \bar{\theta} < \theta_2' \qquad (3.6)$$

In other words, the aggregate reserve-production ratio ($\bar{\theta}$) when total production begins to fall, due to a fall in Q_1, is greater than the critical θs (= θ_1^*, θ_2^*) for the individual deposits. (This result does not depend on $\theta_1^* = \theta_2^*$)

For example, as in the first numerical example above, take $\theta_1^* = 10$, with $R_1 = 150$. Thus $Q_1 =$ starts to fall from a value of 15. Now take $\theta_2^* = 10$, $Q_2 = 15$, and $R_2 = 210$, which means that Q_2 can continue at 15. Furthermore, we have $\bar{\theta} = 12$, or $(150+210)/30$; but even so production will begin to fall (from 30) due to the decrease in Q_1. Algebraically, this is *not* a profound result, but it should lead to some useful thinking about the size of the critical R/Q ratio (= $\bar{\theta}*$) in a multi-field (or multi-deposit) situation. In an important and influential article, Flower (1978) claimed that global calculations should employ a critical R/Q ratio of 15 in order to take into consideration the different sizes and stages of development of producing fields.

The conclusion drawn from all this is that the same phenomenon predicted by Hubbert for the lower 48, and later realized, could happen to the countries outside OPEC – or some sub-set of these countries – *before* R/Q reaches 10. (For example, the decline in the US (of 50 states) began in 1985, with R/Q > 10) The present value of R/Q in these non-OPEC countries is somewhere between 17 and 18, and while many observers are focused on the time remaining before arrival at the magic number 10, it might be useful to think in terms of a somewhat higher figure. It was also mentioned earlier that most oil fields begin their 'decline' before half of their known reserves are exhausted, but in the simple example given above the decline began after 5 years, which was long before the second half. (Note that the 'life' of those reserves was not 10 years (= 150/10), which convention would have, but infinity – remembering, of course, that infinity is not a number, but a direction This might be the reason that Professor Adelman once remarked that oil would be available when we ran out of drinkable water or breathable air.) Figure 3.1 related to this example.

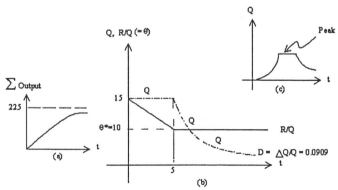

Figure 3.1

Figure 3.1a shows cumulative output, and needs no comment; but it should be understood that the form of Q in Figure 3.1b is not very realistic. Actual fields generally display a form for Q similar to that in Figure 3.1c. Laherrere defines a decline rate for Q after $\theta*$ is reached which can be solved for from an expression of the form $\theta* = (1-D)/D$, but rather than explain the origin of this equation, a derivation will be given that draws on the discussion in this section. Taking two adjacent periods, t and t+1, and using the equality in equation (3.1), we get:

$$Q_t = \frac{R_{t-1}}{1+\theta *} \qquad (and) \quad Q_{t+1} = \frac{R_t}{1+\theta *} = \frac{R_{t-1}-Q_t}{1+\theta *} \qquad (3.7)$$

After a small amount of manipulation we come to:

$$Q_{t+1} = \frac{Q_t(1+\theta*)-Q_t}{1+\theta *} = \frac{\theta * Q_t}{1+\theta *} \qquad (3.8)$$

Dividing both sides by Q_t, and subtracting unity from both sides gives:

$$\frac{Q_{t+1}-Q_t}{Q_t}(=\frac{\Delta Q}{Q}) = -\frac{1}{1+\theta *} = D \qquad (3.9)$$

Notice the use of the delta (Δ) sign: you almost certainly saw it before when you were calculating elasticities in your basic course. Here it stands for $(Q_{t+1} - Q_t)$, and it means 'the change in Q'.

The right hand side in (3.9) is negative, which is as it should be since D is a decline rate; but Laherrere's expression can be obtained by taking the absolute value. In the first numerical example given above, $D = 1/(1+\theta*) = 0.0909 = 9.09\%$. This can be easily checked from the values of Q_6, Q_7, etc that have been calculated.

3.2.1 Exercises

1. Assume that you have 300 units of e.g. oil to begin with, and $\theta* = 10$. Repeat the   example used in this section! Draw the relevant parts of Figure 3.1! What is the decline rate?
2. Suppose that we have 308 units to begin with, but reserves increase by 4%/y, with the increase measured from the amount available at the beginning of the year. Assume that we want to extract 22 units/y for as

Continue the table below until 2 years after production turns down!
Discuss! (Some numbers below are 'rounded').

Year	Beginning R	Ending R	Q	R/Q
0	308	298	22	13.55
1	298	288	22	13.00
2	288			

Repeat the above exercise, but assume that Q is growing at 2%/year, and
reserves increase by 9%/year. Continue the table until you get some idea
of what is happening.

3.   (Extra credit). Show your teacher – and, more important, yourself – how
good you   are with algebra. In calculating the decline rate, assume that
reserves are increasing by g% every year, but everything else is the same.
(In other words, you should take $R_t = R_{t-1}(1+g)$). Prove that under these
circumstances $D = g-(1/1+\theta*)$. Discuss.

3.3   Oil Supply and Demand, and the Reserve-Production Ratio

What I intend to do next is to present a simple derivation which leads to a
surprising conclusion about the R/Q ratio, though perhaps not surprising if
you have done the exercises just above, and in addition have given them
some thought. Consider the arrangement shown in Figure 3.2, which is a two
period (hypothetical) production/reserve scheme, with an annual reserve
growth rate of g (%/y).

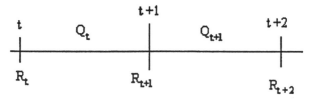

Figure 3.2

Remembering that the rate of growth of reserves is g, we can write R_t –
$Q_t + gR_t = R_{t+1}$, or $R_t (1 + g) - Q_{t+1} = R_{t+2}$. The first relationship above can
be written as:

$$\frac{R_t(1+g)-Q_t}{Q_t} = \frac{R_{t+1}}{Q_{t+1}}\frac{Q_{t+1}}{Q_t} = \theta_{t+1}\frac{Q_{t+1}}{Q_t} \qquad (3.10)$$

What will be assumed next is that the production (driven by demand) grows by n percent per year (n %/y). Introducing $Q_{t+1} = Q_t(1+n)$, and with $R/Q = \theta$, we get:

$$\theta_t(1 + g) - 1 = \theta_{t+1}(1 + n) \tag{3.11}$$

This is a simple first order difference equation, but for readers who do not like difference equations – simple or otherwise – the numerical exercise in the next paragraph should be illuminating. After scrutinizing the exercise, I will solve this equation, and present a result from the calculus. Then, after discussing this result briefly, I will derive something similar to this result – an approximation if you will – using simple algebra, where the only 'sophisticated' notation is the difference operator Δ: e.g. we might have $\Delta Q_t = Q_t - Q_{t-1}$. You should therefore feel free to skip the calculus intrusion!

Assume that at the beginning of a certain period we have $g = 5\%$, $\theta = 10$, $Q_t = 15$, and $R_t = 150$. Now take $n = 0$. With reserves growing, but production constant, we might jump to the conclusion that the reserve-production ratio is growing, but this is not the case. Using the relationships given above, and these figures, we immediately see that $R_{t+1} = 150 + 7.5 - 15 = 142.5$, where 7.5 is the absolute value of the growth of reserves during the period in question. But if Q remains constant at 15 ($= Q_{t+1}$), we end up with $\theta_{t+1} = 9.5$. What has happened here is that output is highly significant relative to the level of reserves, despite what we might think if we take the values of production and reserves at face value, and forgot about certain technical/algebraic realities. Continuing, let us solve the above difference equation (although some readers might, at this point, prefer to move directly to equation (3.15) and the discussion associated with that expression). The solution to that equation is:

$$\theta_t = A\left(\frac{1+g}{1+n}\right)^t + \frac{1}{g-n} \tag{3.12}$$

In (3.12) A is a constant, with $\theta = \theta(T)$ at the time T that we begin our scrutiny of θ. This constant can be easily determined by putting $t = T = 0$; but before doing this I prefer a continuous version of the above equation. This is easily obtained by simply recognizing that $(1+g) \approx e^g$ and $(1+n) \approx e^n$. We thus get as our expression for θ (instead of θ_t):

$$\theta = \left[\theta(T) - \frac{1}{g-n} \right] e^{(g-n)t} + \frac{1}{g-n} \tag{3.13}$$

One of the advantages of the above transformation to a continuous formulation of (3.14) is that it simplifies differentiating that expression. Once we do that, *making sure we remember* that we are interested in θ when $t = T = 0$, we end up with:

$$\frac{d\theta}{dt} = (g-n) \left[\theta(T) - \frac{1}{g-n} \right] = (g-n)\theta(T) - 1 \tag{3.14}$$

An approximation to (3.14) will be derived immediately below, and readers who find (3.14) or the procedure leading up to it mysterious can go directly to those materials; but if you have learned that $d\theta/dt$ means the change in θ with respect to the change in t (time), then you know all that you need for interpreting (3.14). We see immediately that for θ to increase (i.e. $d\theta/dt > 0$), we must have $\theta(T) > 1/(g-n)$, which in the example above is $1/(0.05 - 0) = 20$. Put another way, if $\theta(T) = 10$ in that example, then for θ to increase, reserves must grow by more than 10%: that is, we must have $10 > 1/(g-0)$, or $g > 1/10 = 0.1 = 10\%$.

Now for the astonishingly simple approximation. Remembering that we can write $\theta_{t+1} = \theta_t + \Delta\theta_t$, we get for (3.11):

$$\theta_t (1 + g) - 1 = (\theta_t + \Delta\theta_t) (1 + n) \tag{3.15}$$

or:

$$\theta_t(1 + g) - 1 = \theta_t + \Delta\theta_t + n\theta_t + n\Delta\theta_t \tag{3.16}$$

The approximation consists of the following: dropping $n\Delta\theta$ from (3.16) on the grounds that it is very small, which it is if $\Delta\theta_t$ is small. This, of course, is standard operating procedure in calculus, where that expression would be allowed to approach zero. Now we have:

$$\theta_t [(1 + g) - (1 + n)] - 1 = \theta_t (g - n) - 1 = \Delta\theta_t \qquad (3.17)$$

For all practical purposes, this is the same as the result in (3.14): $\Delta\theta_t > 0$, or θ_t increases, if $\theta_t(g-n) - 1 > 0$. Now let us turn to the real world outside OPEC, where R/Q ($= \theta$) is approximately 17.5, while $n \approx 1.7$. If the reserve-production ratio is to increase, we must have $17.5 > 1/(g - 0.017)$, or $g > 7.4\%$ per year. This may happen from time to time in the future, but it is highly unlikely that a growth rate in reserves of this magnitude could be sustained anywhere in the major oil producing regions outside the Middle East – with the possible exception of the FSU (and with the emphasis on 'possible'. These and other matters are examined with reference to the US by Iledare and Pulsipher (1999).

3.4 A Non-technical Preview of the Oil Futures Market

Now that we have some idea of what is going to happen to the oil market in ten or twenty years, it is time to start thinking about what is going to happen tomorrow, or next week. Needless to say, the ladies and gentlemen who are capable of coming up with the right answers within this particular time frame have some very bright career prospects. Furthermore, if you master the essentials of oil futures (and options and swaps), you are also ready to tackle other markets – gas, electricity, copper, etc; but a word of warning might be in order. The expression 'rocket scientist' originated in finance, and not physics, although the persons so designated prefer to be called 'quants', for quantitative analysts. They and their publicists, and particularly the latter, believe that persons with advanced degrees in mathematics and physics are better prepared to deal with high velocity financial markets than the rest of us. In my case this is undoubtedly true, but in your case it does not have to be true; and in fact the evidence, if carefully examined, will show that it is not true. What you need to succeed in this business is iron concentration; a good knowledge of the fundamentals of derivative markets, so that your mouth does not fall open when you hear expressions such as marking-to-the-market, and the Black-Scholes formula; and a sincere belief that more money is better than less. You might also keep in mind John Kenneth Galbraith's counsel: "Financial genius is a rising market".

Now that the preliminaries are over, a simple example is in order. Suppose that your uncle Pogue phones you from Rome to tell you that all the oil tanker crews in the Gulf are going on strike, but nobody (except Uncle Pogue) knows about it yet. Since you have studied economics, you are able to immediately draw the following conclusion. The supply of oil to the oil importing countries is going to fall, and as a result the price of oil will certainly rise. You therefore pick up the telephone next to your rocking chair, and call your aunt Minnie (who is a commodities broker).

You tell this good woman that you want to buy some *futures contracts* for crude oil. In some circles this would be called buying (or going 'long' in) 'paper barrels', as compared to the physical barrels (or 'wet barrels') aboard the oil tankers that will soon be lying idle. "How many?" inquires Aunt Minnie, and so you tell her what Mr Pogue told you, and then you tell her to use her judgement. She immediately replies that she is going to buy 100,000 barrels for you – which is 100 contracts, since one contract is for 1,000 barrels. She also mentions that she will be buying quite a few contracts for herself. You glance at the latest edition of the *Wall Street Chronicle* and see that the price of *physical oil* is 15 dollars/b, which means that aunt Minnie will be ordering somewhere around 1,500,000 dollars worth of oil in your name. The procedure usually is that she would ask for a deposit – from 5 to 10 percent, which is called margin – but since she knows you, and is familiar with the size of your bank account, she does not bother. Another reason that she does not bother is that she is too busy buying a wagon load of contracts for herself. As far as aunt Minnie is concerned, an opportunity that is erected on the right kind of insider information is the opportunity of a lifetime.

Something which needs to be understood at this point is that you are buying futures and not *forward* contracts. True, a futures contract is a forward contract, in that usually delivery is specified on the contract; but delivery does not have to take place, as you will find out below. Instead, the contract can be offset (i.e. reversed). Forward contracts are very important in the oil business, but the most important thing about them for us is that we can often – but not always – talk about futures contracts as though they were forward contracts, especially in the classroom. It would also be useful to know that futures contracts are *standard contracts*, for a specific amount of a commodity, and should delivery take place because the holder of the contract keeps it to the maturity (or expiry) date, then delivery takes place only to a few specific locations. In our example all futures contracts for oil are for 1,000 barrels, while forward contracts can be for any amount. In addition, if on the spur of the moment you decide to spend a few months skiing in Åre or Courchevel, and forget to offset your contract – or 'reverse' it if you prefer – then the 100,000 barrels of oil would not be delivered to the garden of your humble abode in Bel Air (Los Angeles), but to West Texas or New York Harbor, which are legally specified delivery points.

Continuing, that afternoon when you get home from work, you switch on the television, and hear that tanker crews in the Gulf are indeed going on strike, and already the price of physical oil on the Rotterdam Spot Market has jumped up several dollars, and the same is true of spot oil on the Brent Market (UK). It is being quoted everywhere at $17.5/b, but spokespersons for the oil companies are claiming that everything humanly possible is being done to reach an agreement with the tanker crews. (An aside: oil prices on both the physical and paper markets are extremely volatile, which is one of the reason why risk management (using, e.g. futures and options) is extremely important.)

So you lift the phone again and buzz aunt Minnie. The price of physical oil has gone up, you tell her, and once you read a brilliant book on oil by a great teacher who claimed that when the price of physical oil went up, it was very likely that the price of oil on futures contracts – i.e. paper oil – would also rise.

Yes, she confirms, I know the book, and that gentleman was correct. The price of paper barrels has increased by 2 dollars, but there is an ugly rumor going around that the strike may be settled soon. In addition, the fact that the futures price is below the price of physical oil is not a good sign at the present time, because if you are a (blasé) believer in the *Efficient Markets Hypothesis,* then you might argue that the futures price can be a useful estimate of the spot price in the future. (Observe the expression "useful". What it means is that there are theoretical reasons for not taking the futures price as a perfect estimator of the spot price in the future). Perhaps she should 'close your position' by selling 100 contracts? (Which, as you should note, is the same as the number with which you opened your position).

"Sell now," you tell her. "Dump it all". Everything considered, it was a nice ride, and although it is over you have just increased your fortune (before taxes) by about 200,000 dollars. (If, e.g. the price of oil on futures contracts was now $17/b, you would earn (17–15) x 100,000 = $200,000, assuming that you opened your position by buying 'futures' oil for $15/b). Note too that the price of physical oil did not have to be equal to the price of paper oil when this transaction was initiated, and so when Minnie bought futures, she might have paid less or more than $15. It is possible that you did not ask, however, because with the information you possessed, the price was irrelevant. What was relevant was that the price was almost certain to rise.

Where's the oil? The first time I lectured on futures markets to undergraduates, this was the question of the day. The answer now, as then, is that nobody in Bel Air knows where the oil is, and it is not important. For a speculator the issue is not oil, but profiting on changes in the price of oil.

You picked up a phone, opened a position by buying some contracts, and later you closed your position by selling the same number. You do not worry about where the physical oil is, or where it is going to be – unless you decide to hold your contract to its maturity date. And why would you do that? Answer: because you think that there might be serious money to be made from being in possession of physical oil.

Now let's turn this delightful story around. Uncle Pogue calls from Rome, and tells you that huge new oilfields have just been discovered in Egypt next to existing oilfields. In other words, in order to get the oil to market, no new pipelines will have to be constructed; and since Egypt is having a few serious economic problems, and does not need to abide by OPEC quotas, that oil will be coming to market in a very short time.

Supply up and, *ceteris paribus*, price down! This is what you learned in your first course in economics, and so once again you call aunt Minnie and tell her what uncle Pogue told you. She proceeds to inform you that she is going to *sell* 100 futures contracts in your name – which comes to 100,000 barrels – and she will also make a substantial investment for herself. Some terminology might be useful here. In the previous scenario, when she bought, she went *long*. Now, when she sells, it is called going *short*.

Where's the oil? How can you sell something that you do not have?

If you sell a futures contract, and do not offset it before the maturity date – by buying the same number of contracts (for the same commodity, and having the same maturity) – then you (conventionally) must buy 100,000 barrels of oil, somewhere, and deliver it to a *designated* delivery point. To repeat: just as you opened a position by selling a futures contract, you can close your position by buying a futures contract, and it can all be done from your comfortable living room while you enjoy the stimulating intellectual banter of J.R. and Sue Ellen Ewing on the prospects and personalities of the Texas energy sector. After closing your position, your involvement with the futures market is concluded – at least for the time being.

So much for 'speculation', but what about *hedging* – i.e. some kind of insurance against unpleasant price movements up or down. Suppose that a week ago you came to the conclusion that the oil price was going to rise by a large amount, and so you went down to the local 7-11 and bought 2,000 barrels of oil, which you had transported to the local warehouse for safekeeping until the price went up. But now you are not so sure of the future oil price as you were then, and you feel a strong urge to hedge your investment. How would you initiate this particular risk management exercise?

There are a number of possibilities, but one is to lift the telephone again and ring aunt Minnie. When she answers you tell her to sell two futures contracts (= 2,000 'paper' barrels). Now, if the price of (physical) oil falls, you lose on the 'wet barrels' that you bought, but since the price of futures should also fall (for reasons that are given in Chapter 6), you make a profit when your close your paper position by reversing your initial operation – a short – with a long (which entails buying 2 contracts). The profit is equal to the sale price minus the purchase price. Assuming that you get rid of your physical oil about the same time, the loss on 'actuals' (or the 'underlying' as they are sometimes called) is pretty much counterbalanced by your profit on futures.

It sounds easy, doesn't it, and to a certain extent it is; but never forget that Metallgesellschaft – the 14th largest industrial firm in Germany – lost 1.5 billion dollars in the futures and swaps markets in a few months. (Interestingly enough, when this news came out, the swaps trading of Metallgesellschaft (MG) was completely overlooked). In addition, when a clean-up crew consisting of several world-class traders from the US was brought in to put things right, they apparently made as many mistakes as the people who got MG into deep trouble. Suddenly one recalls a very important Russian adage: The wise man learns from the mistakes of others; the fool must find out for himself.

3.4.1 Exercises

1. In the last exercise above you cancelled your loss on the physical oil you were holding by your gain on the paper transaction. But what would the situation have been had the oil price gone up instead of down? Why might you decide to close (i.e. offset or reverse) your position before the maturity (or expiry) date of your contracts?
2. Suppose that you were sitting in your apartment thinking about the oil that you must buy for your firm next week, and almost frightened out of your wits by the thought of the oil price rising substantially before that time. You pick up the phone and call aunt Minnie, and she suggests hedging your forthcoming purchase with some futures. What exactly does this entail? In your answer, use the term 'short' or 'long', and explain what both of them mean!

3.5 Oil Stocks and Oil Prices

For most conceptual purposes, oil price movements can be categorized as long run (or long term) and short run (or short term). Normally, where commodities such as copper and cocoa are concerned, the long run involves distinct cycles, whose price peaks and troughs are separated by years, or even decades; but this situation does not concern us here for reasons that are obvious to anyone with a serious interest in oil economics: conceivably, the entire modern commercial history of oil will not involve more than a few of these cycles. Instead, we will concentrate on short-term fluctuations where, instead of using the expression 'cycles', we talk in terms of price vibrations or oscillatory movements, whose peaks and troughs are separated by weeks or months.

These fluctuations are the result of speculative tides of bullishness and bearishness, which have their origin in such things as weather forecasts, the outcome of OPEC meetings, and economic forecasts. To many casual observers oil prices appear rather stable, but in reality, even in fairly calm years, sizable price movements take place. Early in the 1990s, peak-to-trough amplitudes of (at least) $3/b were common, and in 1996 these swings increased to about $5/b.

Now let us examine the mechanics behind these movements. What happens is that some relatively minor bullish or bearish changes in perceived supply or demand will often work to accelerate price changes. With an upward or downward trend established on oil price charts – which now can be found in the business pages of many newspapers; and with the economic background characterized by its usual degree of uncertainty, any unexpected new information is quickly given a great deal of publicity, fundamentals (i.e. supply and demand) are hastily reevaluated, and longer-term price implications often tend to be exaggerated. In many cases, this leads to the kind of buying or selling that guarantees a sharp reversal in the existing trend. (And remember: a bull market is rising, while a bear market is one that is declining).

It is here that we can introduce an interesting observation from the Centre for Global Energy Studies that was published in mid-1993: "The consensus view is that prices will rise in the fourth quarter due to a rising call on OPEC oil, but this view is too simplistic: the market will continue to be driven, as always, by the industry's desire to hold stocks." Among other things, this means that we do not need to be unduly concerned with the psychology of individual market transactors, and can concentrate on the hypothesis that oil prices are highly correlated with carry-over inventory levels, which is basically what the *supply of storage theory* is all about. On both the theoretical and empirical level we observe that inventory levels can be linked to spreads between the prices on various maturities of futures contracts, or even spreads between spot and futures (and/or forward) prices. (As an example of various maturities we could have 30 day, 60 day, 90 day, etc contracts. There is, regrettably, no continuum of futures prices, which tends to make real world futures markets very different from those we find in the learned journals, if we desire to argue the efficiency of the price system under uncertainty; however an interesting use of actual futures markets for estimating oil properties evaluation can be found in Pickles (1994).)

A few months later the market for spot (or prompt) crude was still comparatively weak, but not nearly as weak as around mid-year. What happened was that, in the late summer of 1993, the market was sold into *contango* which means that the spot price of oil was less than the forward price (which in most markets is considered the normal situation) by an amount which exceeded the storage cost. This made it attractive for holders of physical oil to buy spot and sell forward, because for all practical purposes the downside risk was eliminated. Furthermore, hedging transactions were feasible all the time during this period which involved the purchase of spot crude (as in the preceding section) and the shorting of futures contracts: if the price of crude fell, a compensating profit would be realized on the offsetting futures transaction. Needless to say, a situation where oil can be stored without risk, even in a falling market, greatly increases the incentive to hold stocks. (But remember that many stocks are not hedged with futures or forwards, because their owners believe that superior profits are possible if they are left unhedged).

Several months later the contango substantially decreased, and with the price of futures contracts also falling, the demand for all categories of stocks fell drastically. A major part of new production, together with a large amount of existing stocks, were then directed into the spot market, and this eventually depressed the spot price to about $14.75/b. But as a result of the OPEC meeting of 25-29 September, which set the overall production ceiling of OPEC at 24.5 Mb/d, the spot and forward prices recovered in such a way that destocking ceased. Together with growing OPEC solidarity, which means an increasing willingness to address the problem of 'production indiscipline', and given expectations of a satisfactory macroeconomic development in most of the world, particularly Asia, the forecasts were for an acceptable level of price. This brings us to Figure 3.3.

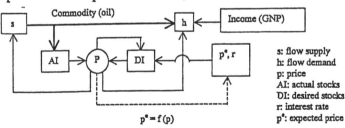

Figure 3.3

Figure 3.3 is a stock-flow model of a commodity market. It applies to any market in which inventories (i.e. stocks) are the key item in determining short-run prices, which is the case in the crude oil market. This construction implies an inherent volatility of the oil price – a volatility that has made it a paradise for speculators; and as made explicit in the model, as well as in the preceding discussion, inventory behavior is very heavily influenced by changes in expectations. Readers should thus be especially aware that since prices on futures markets are often taken as the best estimate of spot prices in the future, these markets encourage the frequent overhauling of expectations. This helps to promote instability. (I will also mention once more that there are theoretical arguments which suggest that the futures price is not a completely satisfactory estimate of the spot price in the future).

Also note that the arrangement shown in the above diagram is analogous to a servo-mechanism, whose purpose is to regulate AI in the light of DI, comparing these two and translating the discrepency (if one exists) into an 'error' signal (DI – AI), which in turn acts on flow supply and demand to eliminate this discrepency.

If the mathematical equations (i.e. differential equations) of this sytem were written out in full, and included the case in which p^e was a function of p, as shown by the dashed line in Figure 3.3, then when they were solved we would find a greatly expanded scope for instability, as compared with the case where p^e is independent of p. Eventually, some readers might find it challenging to undertake this project.

That brings us to the next level of difficulty in relating stocks and the price of oil, with our topics here being the extremely important *supply of storage* and *convenience yield.* First we turn to Figure 3.4a, which shows the (months) of crude oil inventory cover prevailing at a certain time, plotted against the percentage price spread per month, between the present month and a future month (e.g. February 2000 and May 2000). Strictly speaking, this is the correct representation, while the arrangement shown in figure 3.4b is an approximation; but it is one of the approximations that can sometimes greatly assist us in comprehending a very complex subject.

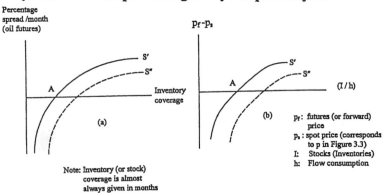

Figure 3.4

What these curves indicate is that to the left of point A, we have the condition known as *backwardation,* with the price of the nearest futures contract greater than that of the 'next nearest' contract, or, for that matter, contracts of longer maturities. (An even easier example would involve the spot price of a commodity at time t being greater than the futures price for time (t+1.) We obtain this result when inventory coverage is relatively low, and oil users and traders believe that they may be able to profit from a shortage of oil by holding inventories, or titles to inventories, rather than other assets. As will be shown later, there is a convenience yield being realized here which militates against selling the commodity (oil) and replacing it with a futures contract (on which delivery could be made). In addition, the cost of ordering and transporting commodities from one location to another, and the need to maintain production schedules, justifies the holding of some stocks in the face of the strongest backwardation.

Having considered backwardation, we can look at the situation to the right of point A, where inventory coverage is more than adequate, and additional stocks have a comparatively small (marginal) convenience yield, or in the limiting case hardly any (marginal) convenience yield at all. (Note the introduction of the term 'marginal'. Obviously, when we are far to the right of A, there is hardly any addition to the convenience yield due to the addition of an extra unit of inventory – the marginal convenience yield approaches zero; but the average or total convenience yield should remain the same.) Accordingly, once we are far to the right of A, additional stocks will be carried by profit maximizing transactors only if they expect to recover the storage and transaction costs involved via an imminent rise in the price of oil, or through being able to hedge stocks by selling futures (or forward) contracts. Otherwise, these (excess) stocks would be put on the market, and on the basis of Figure 3-4b we can reason that this would *ceteris paribus* depress the spot price p_s. This would establish a positive increments between the spot price and the futures contract with the lowest maturity – and likely between successive futures contracts of higher maturities – that are in excess of carrying costs, which in turn would induce inventory holders to carry the remaining stocks. Note also that to the right of A we have the condition called *contango*.

In Chapter 9 some elementary algebra is employed to discuss the above topics, but in truth a few well chosen words will take us a long way toward the understanding we need. Inventories almost always provide 'convenience', which is often labled a convenience yield; however the *marginal* (i.e. additional) convenience yield approaches zero for very high inventory levels. Similarly, a high marginal convenience yield is thought to characterize a situation where stocks are considerably below adequate or normal levels. Furthermore, in the latter circumstances, a large marginal convenience yield can cause the futures price to be lower than the spot price. As our algebra makes clear, a marginal convenience yield is a *negative* cost, and signifies the increased utility to the inventory holder of an additional unit of a commodity that can be sold or traded at a later date, or included in the production process. Apparently there is some controversy concerning the convenience yield in general, with some investigators claiming that it is a chimera. This may well be so for some commodities, but oil is definitely not one of those commodities.

As is customary in economics, we need to designate an equilibrium – since our economics is, specifically, equilibrium economics: we study equilibria, and only very seldom the movement between equilibria. The most general definition of an equilibrium (in economics) is a condition where expectations are correct or fulfilled – or, to be more precise, where the outcomes of transactors' efforts turn out to be consistent with their intentions. In terms of Figure 3.3, for example, our equilibrium would be when we had AI = DI, and transactors would no longer attempt to change AI. The definition of equilibrium that is sometimes employed in physics might also be of use here: a state of rest. (*And reader, make it your business to master the diagrams in this section. They are extremely important!*)

This immediately leads to a consideration of scenarios that result in disequilibrium situations. One of these might be where the holders of non-hedged or only partially hedged stocks revise their expectations, and sell futures contracts in order to hedge their stocks. This will lower the price of these contracts (since it raises their supply), and therefore will reduce the attractiveness of holding inventories for existing hedgers, who must periodically renew contracts. (For reasons that cannot be gone into in great detail at the present time, short-term futures contracts that are continually 'rolled over' can sometimes be more attractive than fairly long term contracts. This was Metallgesellschaft's hedging strategy, although in their case it brought them loads of grief. The logic here turns on the comparative shortage of liquidity for some contracts: e.g. oil futures whose maturity is longer than one year, and sometimes less.)

As discussed above , we now get considerable destocking, which leads to a strong downward pressure on the spot price of physical oil. Assuming no futher change in expectations dealing with physical oil, a price spread between the spot and futures (and forward) price would now come about which would make it attractive for inventory holders to cease destocking.

The most important point of all is that, over a short period of time, stocks are not determined by economics, but by the physics of production and consumption. In the short-term inventories are what they are, and when their owners come to the conclusion that they are too large, the only favorable adjustment possible – assuming that their use in current consumption and production activities cannot be increased by a substantial amount – is that the price of futures contracts changes in such a way as to provide the compensation for holding what have come to be regarded as excessive stocks. This is asking for a great deal, since the spot price of oil will almost certainly decline as excess stocks are unloaded (i.e. redistributed), and /or the attempt to hedge what are regarded as exorbitant inventories would lead to a fall in the price of oil futures. As for the strategy to make money out of all this, the most important thing for a speculator is to start selling futures just before everyone else discovers that inventories are too high. For example, you didn't have to be a weatherman to know which way the wind was going to blow when, as in the late 1990s, OPEC unexpectedly raised its quotas. In that situation only one thing could happen, and it did: the oil price began moving South, and continued to do so until oil producers recovered their senses. (And oil producers here included those in non-OPEC countries who decided to cooperate with OPEC.)

As some readers have probably guessed, a fall in the demand for inventories could be associated with an upward shift in the supply of storage curves shown in Figure 3.4. The reader will be left to mull over this, although the opposite arrangement – an exogenous increase in the demand for stocks – will be discussed below. But first, let us say something about the shape of these curves. The right hand side of the curve should be relatively flat, because with adequate inventories, and more or less constant carrying costs, spreads do not need to increase so that they are greatly in excess of normal carrying costs. In this case the *marginal* convenience yield approaches zero. Similarly, when I/h is low, the marginal convenience yield is high, which among other things reflects a rapidly increasing financial penalty for not selling this oil, and replacing it by a futures contract (on which delivery could take place to a favorable location). Geometrically, this implies a very steep portion for this phase of the supply of storage curve.

Now let us look at a situation where transactors conclude that existing stocks are inadequate, which in Figure 3.4 calls for a downward shift of the supply of storage curve from S' to S". Once again we recall that instantaneous additions to stocks cannot take place, and before production is increased, transactors bid for existing stocks, thereby raising the spot price of oil. This arrangement is shown in Figure 3.5. (Notice the units: barrels (b) for stocks of e.g. oil, and b/time-period for flows.)

The story here is as follows. The increase in demand for stocks is shown by the movement from D' to D" in the stock market, which causes the price to increase to p". In the flow market, when the price increases to p", flow supply exceeds flow demand, or s > h: more is being produced than in being consumed in that period. This causes an increase in the level of stocks in that period, and in the diagram on the left, the stock 'supply' curve, I', moves to the right by an equal amount.

In the next period, if inventories (I) have not reached I* (which is the case in the diagram), then we still have p > p', and so once again s > h. There is a further increase in stocks which, judging by the curves that have been drawn, is less than the original ΔI. As the level of inventories increases, demand for new inventories slackens, which explains the fall in price from p" toward – and eventually to – p', where once again s = h.

This story can also be told with Figure 3.3. When I increases from I' to I* in Figure 3.5, we have DI > AI. This influences p in such a way that s increases, and h decreases. As DI falls, p falls, and so as above we eventually get AI = DI, and s = h. What we need to note here is that the centerpiece of these models is the stock market, and not – as in your first course in economics – the flow market. Of course, this smooth chain of events could be frustrated if p and the change in p with respect to t (Δp/Δt) influences DI in the 'wrong sort' of way, causing the new equilibrium, while definable, to be unattainable.

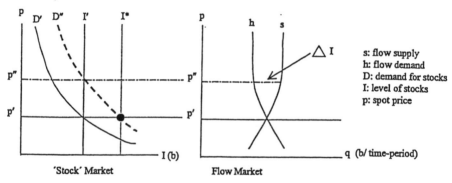

s: flow supply
h: flow demand
D: demand for stocks
I: level of stocks
p: spot price

Figure 3.5

3.5.1 Exercises

1. A simplification was used to get equation (3.19). Go back through the algebra and do not use this simplification. Would it influence the argument we were trying to make? Discuss!
2. Speculating with futures is fairly straightforward, as we saw in the soap opera involving Uncle Pogue and Aunt Minnie; but hedging with futures is not always appetizing to some producers. They say that it doesn't leave any room to take advantage of favorable price movements. What could they mean?

3. In Figure 3.5 we saw and discussed what happens when the demand for inventories increases. Show and discuss what happens when this demand decreases. What happens on the futures markets in this situation?

3.6 Conclusions

The purpose of this chapter is to give all readers some insight into how the world oil market functions. Please let me emphasize however, that regardless of appearances, it does not contain a *model* of that market. It contains parts of a model. The complete model – correctly specified and ready to tell us all that we want to know – will probably never be built.

Still, there are many other parts or components that I would have liked to take up. For example, the ability of market prices to transmit the kind of information we tell our students they are capable of transmitting. Take the case of the Cerro Azul Number 4, located near Tampico, Mexico. When it 'blew' in on 10 February, 1916, a column of oil rose 600 feet in the air. Its initial production rate was 260,000 b/d, making it the most productive oil well in history – for a while.

After producing 60 million barrels, it suddenly produced nothing but salt water. The kind of price systems that we take such pleasure in discussing in microeconomic theory, on any level, cannot deal with this kind of phenomenon – except at the Alpine heights of the very purest theory, far away from where real people live their lives. Certainly, nowhere in (or near) the first course in economics did readers come face-to-face with a highly productive activity that ended with a 'bang' rather than a 'whimper'. In financial economics we say that no one individual is bigger than the market, but in energy economics the market must often bow to geology,

Virtually every year we are told about new oil ventures that promise unfathomable riches. As I write these lines the new El-Dorado is supposed to be Central Asia. In 1997 it was the Hibernia field, about 310 kilometers due east of St. John, Newfoundland (Canada) in a part of the North Atlantic known as 'iceberg alley'. 4.2 billion dollars were initially earmarked for this project, since ostensibly Hibernia's two reservoirs contain a minimum of 615 Mb of oil, with 'perhaps' another billion barrels in other fields nearby. Deep water exploration and drilling is apparently to play a major role on the international oil scene in the 21st century, although a "giant oil play" has been trumpeted in the Italian Apennies by the present director of the Italian oil giant ENI, founded in 1953 by the legendary Enrico Mattei, whose brand of economic nationalism may have led to a price being put on his head by certain rivals and detractors.

What is apparently being missed is that Hibernia type projects could only be justified economically if certain persons in certain executive suites expect oil to take a definitive move up the price ladder in the not too distant future; while the talk of an oil bonanza somewhere beneath the ski trails of the Apennies is mostly directed toward the Italian share market, where the truth is sometimes defined as anything that cannot be proved a lie.

All readers will not find the story being told in this conclusion congenial. For those who want other points of view, let me suggest the important non-technical discussions often published in the (quarterly) IAEE Newsletter.

3.6.1 Exercises

1. Open interest is the number of futures contracts still open at the end of each trading day, counting either buy contracts (longs) or sell contracts (shorts) – but not both. Why is open interest such a good indicator of the liquidity of a contract?
2. Michael Lynch (1999) and Paul Stevens (1995) might not agree with my arguments about an impending problem with the supply of oil. Examine their work and comment.
3. M. King Hubbert said that production in the US lower 48 would peak between 1965 and 1970. It peaked in 1970. Is there a possibility that King was just 'lucky'? Examine the article of Cutler Cleveland and Robert K. Kaufman (1993) for a detailed analysis of this matter.
4. If futures contracts are so wonderful where the obtaining of price insurance is concerned, then why don't the oil producing countries use them?
5. There was a lot of talk in this chapter about stock-flow models. Given an example of a pure *flow* market , and a pure *stock* market.
6. Arguments are often given in elementary textbooks that the presence of forward or futures markets will reduce price volatility for various items.Why? But the opposite might also happen. (The appearance of a futures market does not appear to have reduced price volatility in the oil market.) Discuss this matter, using if possible Figure 3.3.
7. I say at one point in this chapter that there is no credible *microeconomic* theory of technical change. Comment! But textbooks in economics have a great deal to say about technical/technological change. Have I made a mistake?

3.7 APPENDIX: THE DOMESTIC INDUSTRIALIZATION OF HYDROCARBONS

"The domestic industrialization of hydrocarbons" is a phrase used by Alvaro Silva, the deputy minister of energy of Venezuela. What he means is that for Venezuela and other hydrocarbon rich countries, an expanded petrochemical (and refining) industry can generate increased employment, attractive returns for investors, and important tax income. Going somewhat further, it is an ideal industry to serve as one of the motors of economic growth.

Venezuela is a country with abundant reserves of inexpensive natural gas that can serve as a feedstock for the petrochemical industry, and the same is true of the augmented and broadened stream of refining products that are under consideration. This kind of situation is duplicated in much of the Middle East, since Saudi Arabia, the United Arab Emirates, Qatar, and especially Iran, each have more confirmed gas reserves than the whole of South and Central America, to include the Caribbean region. Iran, of course, is a potential gas 'giant',

It is also interesting to note that an improved gas-to-liquid technology will eventually be capable of producing large amounts of relatively clean liquid fuels that are lower in sulphur and other pollutants than conventional fuels.

The main outputs of a petrochemical industry are aromatics, ammonia, methanol, and olefins. The products obtained from these are inputs for activities yielding plastics, paints, fibre/textiles, pesticides, fertilizer, pharmaceuticals, etc. There is no point in making grandiose predictions about an industry where competition is as stiff, and technology as important, as in this industry, but if 'comparative advantage' still has any meaning, then the Middle East is very well situated.

By way of completion, something should be said about refining. John D. Rockefeller – the founder of Standard Oil – was of the opinion that pumping crude oil was a waste of time: the real money was in refining. The usual refinery functions as follows. Crude oil is pumped into a distillation tower that is pressurized, and is hotter at the bottom than at the top. The various oil products have different boiling points, with those that are the lightest having the lowest. When crude oil enters the tower, the heaviest part remains in liquid form and falls to the bottom. The rest is vaporized, but the various constituent products return to liquid form as they reach the lower temperatures higher up the column. As a result they can be piped away.

Refinery products are usually divided into three cuts or fractions: gas and gasoline (petrol), or 'light products'; middle distillates; and fuel oil and residual cuts. Much – but not all – of the time the most valuable end of the barrel is the top end, which provides the so-called white products: domestic gases, aviation fuels, motor fuels, and some feedstocks for the petrochemical industry. Naptha, which is important for the improvement of gasoline quality and is also a valuable petrochemical input, is extracted from both the light and middle ranges of distillate cuts. The other middle distillates are kerosene, paraffin, and light gas-oil. The rest of the refinery output consists of heavy lubrication oils and residue.

This short comment can be concluded by assuring the reader that the refining industry is one of the most competitive in existence, and the many independent refiners throughout the world – i.e. those without an assured source of crude – are accustomed to thinking of themselves as an endangered species. As mentioned earlier, refineries are an important source of petrochemical inputs, to include the fuel to operate petrochemical facilities, and in looking at the entire package, it becomes even more obvious that the global center-of-gravity of industrialized hydrocarbons could eventually be located in the Gulf.

Chapter #4

A Fuel of the Future: Natural Gas

This chapter presents a brief, up-to-date analysis of the world natural gas market. Along with a sketch of supply, demand, and price, some consideration will be given to the deregulation-privatization controversy. The most striking thing about this issue is that it does not deal with deregulation as such, but with reregulation: the changing of rules. Similarly, the question is not privatization, since most gas companies are already in private hands, but 'fragmentization'.

Unfortunately, it is almost impossible to do justice to the many facets of natural gas in a single chapter, particularly when we see that natural gas is the fastest growing major source of energy, and has been since 1995. But on the other hand, the gas market may not require the same kind of attention as the oil market, where a rapid price escalation could also have far-reaching macroeconomic consequences. The estimated global consumption was about 80 Tcf in 1999, which means an 8% total growth rate over the 3 preceding years; but with consumption growing for all fuels, the worldwide gas market share should remain in the vicinity of 23 percent.

4.1 Geology, Units, and some Economics

Natural gas is usually found in an environment similar to crude oil, and at one time was frequently called gaseous petroleum. (Petroleum is oil *or* gas). The hydrocarbons of natural gas are lighter and less complex that those of crude oil, and natural gas occasionally contains water and substances that are not hydrocarbons. Like oil, gas is found in water-coated pore spaces in rocks – for the most part sedimentary rock classified as organic shale. This shale originated as the remains of prehistoric plants and animals, and was 'cooked' into oil and gas by heat, the pressure of earth acting over millions of years, and various chemical reactions.

Some hydrocarbon deposits contain oil but no gas, while others – where the cooking referred to above continues until hydrocarbons were reduced from liquid oil to molecules of gas – contain gas but no oil. This latter category is called non-associated gas. A common arrangement is the presence of gas and oil in the same deposit, and in this situation the gas is called associated gas. Since non-associated gas may come into contact with water, but not oil, its production is discretionary: it can be 'shut in' (i.e. left undeveloped) until market conditions warrant its exploitation. By way of contrast, associated gas is not discretionary because it becomes available whenever the associated crude oil is produced, and thus if it is not piped away to be sold it can be flared (i.e. burned up in the air), or reinjected, where it helps maintain reservoir pressure. Nigeria once flared almost 1.5 Gcf/d, equal to 3% of US demand at the time.

In 1983 about 6 percent of total world gas production was flared, which was roughly 106 billion cubic meters (= 106 Gcm); and that was probably a maximum for the percentage of gas disposed in this manner. Most of the flaring took place in the Middle East and Africa, where the largest amount of associated gas is located. But since that time large gas-gathering systems have been constructed to capture this associated gas, and flaring may have been reduced to about half of the above percentage.

A large source of losses can be the processing activities associated with liquefaction. Similarly, reinjection come to about 5% of gross production, but here it should be observed that reinjection often permits natural gas producers to delay their decision on the use of associated gas, since up to 80% of this reinjected gas can be recovered after oil production ceases, depending on the nature of the deposit. It is believed that gas can be found at much greater depths than oil, and new drilling technologies are being developed that will permit probes below 30,000 feet. It has been suggested that once exploration is fully underway at these depths, world reserves of gas might show a sizable increase. A similar observation applies to offshore gas, particularly in very deep water.

Before continuing, a few things must be said about units or equivalencies (and here the reader can refer again to Chapter 1.) Thermally, or in terms of heating values, 1,000 cubic feet (= 1,000 cf) of natural gas is equivalent of 0.178 barrels of crude oil, *on the average*. In addition, 1,000 cubic feet is equal to 28.3 cubic meters (28.3 cm), or 1 cubic meter = 35 cubic feet. Thus 1,000 cubic meters of natural gas is approximately the equivalent of $(1000/28.3) \times 0.178 = 6.3$ b of crude oil. It is also useful to know that one cubic foot of natural gas has an average heating value of 1,035 Btu, which in turn is equal to 1.055 million joules. We usually introduce an approximation here, however, with 1,000 cubic feet of natural gas being assigned a heating value of 1 million Btu. (Note the approximation here: 1 cf = 1,000 Btu instead of the correct 1 cf = 1,035 Btu.)

Price is often quoted in terms of million Btu – for example, $2.50/MBtu. In 1979-80 this was the average price of gas in the United States, which in oil terms is the equivalent of about $14.5/b, and since on average oil was selling for $30 per barrel, gas was a very popular medium. In 1979, just before the second Oil Price Shock, Finland bought 993 Mcm of gas from the FSU for $61M; but in 1980 imports dropped to 925 Mcm, although the total price was $111 million (=$111M). The Finns were not happy with this price increase, although relative to the world price of energy it was not so bad. 925 Mcm for $111M gives a price of $0.12/cm of gas. Multiplying this by 28.3 (cm/thousand cubic feet) gives a price of $3.396 per thousand cubic feet or, approximately, $3.396/Mbtu. Since 1,000 cf of natural gas is the equivalent of 0.178 barrels of oil, on the average, the equivalent oil price is then 3.396/0.178 = $19/b. This was more than the previous price, but at that time still under the *thermal parity* (with oil).

We can now return to our main theme. Conventional non-associated natural gas from a well consists mainly of methane – on the average about 85%; other hydrocarbons known as natural gas liquids (produced together with natural gas, but liquid under normal conditions); water, carbon-dioxide, nitrogen and some other non-hydrocarbons. On the other hand, associated gas may contain between 20 and 50 percent of gas streams other then methane. Methane can be thought of as pure natural gas, while the other gas streams are highly suitable as petrochemical feedstocks, and fuels and inputs for refinery processes. Figure 4.1 shows the components of natural gas, ex-well.

Many energy buffs are now working overtime with their crystal balls and other equipment in an effort to determine just what surprises are awaiting us in the 21st century. On the positive side, the idea of a recycle society seems to have taken hold. In Figure 4.1 we can see that when we are talking about natural gas, we are mostly talking about methane. Thus we can immediately ask if there are not other usable sources of methane than those pictured in this diagram. Business Week (July, 1999) thinks that a great deal of the energy used in the home will be generated in the home. Homeowners, for example, will be able to recycle garbage, human waste, and other effluvia to obtain methane rich gas that will work as an input in generating electricity and producing heat. Not only that, some of this home-grown electricity can be sold to firms.

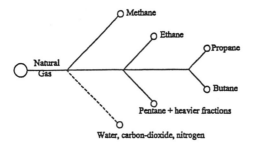

	Liquified at	Definitions
Butane (normal)	-0.5°C	NGL: Natural gas liquids
Butane (Iso)	-12.0°C	LPG: Liquified petroleum
Propane	- 42.0°C	LNG: Liquified natural gas
Ethane	- 88.0°C	SNG: Synthetic (substitute) natural gas
Methane	- 161.0°C	

Note: Natural gas = methane + NGL + (water, nitrogen, CO2).

NGL = ethane + LPG + (pentane and heavier fractions).

Lpg = propane + butane + mixtures of propane and butane

Figure 4.1

Before dry natural gas is distributed to consumers, undesirable components are removed and an attempt is made to obtain a uniform quality by decreasing the share of hydrocarbons. Natural gas liquids are separated out, and where a market exists the more valuable components – butane and propane (i.e. LPG) – are sold, usually under the name gasol or bottled gas. In those countries where the availability of natural gas is greater then the absorption capacity of local markets, and where it is not possible to transport the gas to foreign markets in its original form, some uncomplicated further processing of natural gas can be a valuable commercial activity.

Initially the gas industry was based on town gas (or manufactured gas), which is not natural gas but a gas manufactured by carbonising coal. This fuel was first introduced to the larger cities of the world in 1812 (in England), and is still used in the German Ruhr. The United States first used town gas in 1816, and 5 years later the first use of natural gas was recorded at Fredonia, New York. The first long distance all-welded pipeline (of 14 to 18 inches and 217 miles long) was put into operation between Louisiana and Texas in 1925, and this can probably be taken as the beginning of the modern gas industry. However Carr (1978) feels that "the age of natural gas" was born about 1935, when thin walled pipe could be successfully welded to yield long pipelines of sufficient strength to carry gas under extremely high pressures.

The most 'natural' way of moving gas is via pipelines, at least up to distances of about 8,000 kilometers. In North America and Western Europe, pipelines can distribute natural gas with almost the same efficiency and flexibility as national and local grids can distribute electricity. In the US this sometimes led to intense competition between gas and electricity in various final consumer markets. Like electricity, gas can also be distributed continuously and delivered at precise rates to households and industries.

Gas pipelines are not inexpensive, however. In the 21st century the Black Sea Countries (e.g. Turkey and the Ukraine) seem destined to act as a bridge between the enormous gas reserves of the Middle East (to possibly include Turkmenistan) and the huge European market. One proposed pipeline is a joint venture between Gazprom of Russia and ENI of Italy, which is called Blue Stream, and which would be the world's 'deepest' pipeline: as envisioned, it would plunge to a depth of 2,150 meters in the Black Sea. (Of course, this may turn out to be a 'pipe dream'.) The expected total cost of a project of this nature has been put at 4 billion (US) dollars.

Another option is to cool gas to −160 degrees (−160°) Centigrade by a cryogenic cooling process. This reduces the volume of the gas by a factor of 600, and from the liquefaction plant the liquefied natural gas (LNG) is transferred to a cryogenic carrier and transported to the consuming country, where it is regasified. Some of these ships can carry an amount of gas that is the equivalent to more than 450,000 barrels of oil. It is sometimes claimed that LNG trade is more flexible than piped natural gas because LNG carriers can be rerouted, while pipelines are fixed. In considering the long-term contracts under which most gas is sold, along with the comprehensive pipeline networks that can be found in Europe and North America, I doubt whether this claim is valid. One thing that should be understood is that transporting gas in any form is more expensive than transporting an equivalent amount of oil (measured in heating or calorific value) by a comparable carrier. Something else worth keeping in mind is that the price of LNG, in dollars/Mbtu, is almost always appreciably higher than pipeline gas.

Natural gas production takes place both on and offshore. Offshore production has grown very rapidly since 1980, and by the end of the 20th century amounted to at least 20% of all commercial production. The United States accounts for almost 50% of world offshore production, and about 20% of known US gas reserves are offshore. Western Europe also has sizable offshore production, mostly in the Norwegian and UK North Sea. It is believed that the Arctic waters offshore from Norway and FSU contain large quantities of gas, but this gas will almost certainly be costly. Furthermore, it seems very unlikely that private firms would be willing to search for and produce this gas unless they could 'hedge' their price uncertainty with very long-term 'take-or-pay' contracts. There is too much to lose on both the production and the marketing side for them to be willing to trust their fortunes to short term (e.g. spot market) arrangements. I have the same opinion about the electricity market, where a fairly large percentage of newly ordered power stations supposedly falls in the so-called 'merchant plant' category, whose entire output is intended for the spot market. *Ex-post*, however, an entirely different arrangement may prevail.

This might also be a good point to say something about the large and extremely important petrochemical industry. About two-thirds of the energy consumed by this industry is in the form of feedstocks for the many products it produces, while the other one-third supplies the direct energy inputs needed for various processes. This last statement explains why petrochemical industries constructed in the Middle East are doing so well: very low cost natural gas, and perhaps some oil, supply the fuel needed to 'drive' their production activities; while refined products (such as naphtha and ammonia) manufactured from low cost oil and gas serve as feedstocks.

Among the outputs relevant here are such things as plastics, detergents, nylon, pesticides, pharmaceuticals, and disinfectants. Using gas as an input in ammonia production is especially meaningful, since ammonia is a key ingredient of nitrogen fertilizers. If domestic ammonia production is large enough to realize significant economies of scale, developing countries with gas resources might be able to produce large amounts of fertilizer, and thus avoid costly imports. Unfortunately, very few persons seem to realize the importance of agricultural chemicals, and what a large rise in their prices could mean for the living standards of many hundreds of millions of people – especially in the Third World.

As alluded to earlier, a substantial increase in the price of oil and/or gas will raise the cost of refinery inputs, and thus lower the competitiveness of many refineries (and petrochemical plants). Early in the 1980s the Taiwanese petrochemical industry was in serious difficulties because of rising energy prices. When these prices fell, industry managers breathed a deep sigh of relief, but also intensified their interest in nuclear energy and, to a certain extent, coal, since coal was regarded as the fossil fuel that would be the least expensive in the long run. The nuclear energy would be used for the non-feedstock component of the energy input, as would coal, although it was recognized that the day might eventually come when feedstocks that are now manufactured from oil and gas would be manufactured from coal.

The average rate of increase in world trade involving gas was slightly in excess of 4% during the 1990s. The largest increase was LNG trade, although the leading exporter was the FSU, whose exports were carried exclusively in its large diameter pipelines. These pipelines, incidentally, should have been even larger, and would have been had it not been for the attempt by the government of former US president Ronald Reagan to boycott gas from the FSU.

The largest exporter of LNG is Indonesia, with most of its exports going to Japan and South Korea. Other large exporters of LNG are Malaysia,

Brunai, Australia, Abu Dhabi, and Libya. The United States is a fairly large gas exporter, particularly with pipeline gas to Canada and Mexico, and LNG to Japan; but at the same time that country is a very large buyer of both pipeline gas and LNG.

One of the most interesting things about the natural gas industry is that it has extremely valuable lessons to teach both energy and development economists. Opening borders wider was supposed to be a good thing for all parties, but another picture emerges when we examine the US-Canadian gas trade. Gas prices in Canada have been 30-50% lower than those in the US, due to intense competition in the Canadian market. But new pipelines added 1.1 billion cubic feet (= 1.1 Gcf) – or about 15% – to Canadian export capacity as the 20th century was coming to an end, which raised spot prices in Canada by about 30-40 percent in a year, and reduced the traditional differential between Canadian and US gas prices by approximately 75%. Long term contracts for the winter of 2000 were in some cases selling for over $2.00/MBtu, which led to increased gas production, but also increased stockpiling for future deliveries. As per the stock-flow model in Chapter 3, increased prices in Canada were unavoidable.

On the other hand, although only possessing 0.3% of world confirmed gas reserves, the small Caribbean state of Trinidad and Tobago seems ready to become an important player in the world LNG market. The government of that country has finally decided to make the most of its energy reserves, and has signed long term 'take-or-pay' contracts of up to 20 years for the expanding output of its LNG facilities. Plans also have been forwarded for raising the already considerable production of nitrogenous fertilizers and methanol, and perhaps to begin the construction of an ethylene complex. These capital intensive and energy-based activities, together with an expanded tourism, are capable of invigorating the economy of Trinidad and Tobago.

Finally, an observation about gas prices. In the European Union (EU), during the first part of the 1990s, the price of LNG averaged about $2.5/Mbtu, as compared to $1.75/Mbtu in the US. Roughly, during the decade 1985-95 the price of LNG was one dollar more per MBtu than pipeline gas.

4.2 Economic Theory and Natural Gas: An Introduction

This fairly long section will deal with 3 aspects of natural gas economics: production, transmission, and storage. Much more, perhaps, should be done here, but even in my book on natural gas (1987) it was impossible to consider all facets of this very sophisticated industry. Some idea of its sophistication and extent can be obtained from Figure 4.2.

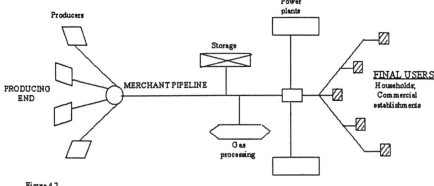

Figure 4.2

One of the most interesting aspects of modern gas history is the lumpy nature of the gas discovery process, beginning with the discovery of 5 supergiant fields in the 1920s, in Texas and New Mexico. The next great period of discovery began in the 1950s, and peaked in the 1970s. Among the bonanzas uncovered was the Hassi R'mel field in Algeria, which has become so important for Europe due to an underwater pipeline from the North African coast to Italy. There were also exceedingly rich strikes in Libya, Anu Dhabi, Saudi Arabia, Nigeria, and France (the Lacq field).

For Europe, however, the great find was the Slochteren field in Groningen Province, Holland, in 1959. This helped to spur prospecting that led to the extremely valuable North Sea discoveries, beginning in 1965 with West Sole in the UK North Sea, and eventually proceeding up to Norway's Troll Field, which ostensibly will still be in full operation through the first quarter of the 21st century, and might produce substantial quantities after 2025.

Once the expansion of the gas network in Western Europe began to accelerate, it became impossible to ignore the huge reserves of the FSU, which is the most gas rich region in the world. It is useful to note that while the FSU oil sector has fallen on bad times, the gas sector in those countries seems to be operating fairly smoothly. Furthermore, it has been suggested that the construction of the Urengoy pipeline from Western Russia to Western Europe demonstrated that there are occasions when commerce can be stronger than politics, since although the cold war was raging at full blast, the communists fulfilled their side of a complicated deal with capitalists to the letter.

Just as Russia holds the lion's share of gas reserves in Europe, Iran occupies this position in the Middle East, and eventually could prove to be a major supplier to Europe. At the same time it should not be forgotten that there is a major market for gas in the Middle East itself. This involves gas for generating electricity, desalinating water, household uses, and as a fuel and feedstock for basic metal industries, refining, petrochemicals, etc. Gas might also turn out to be an invaluable input for the manufacture of vehicle and aircraft fuels, since it appears that it is now possible to turn gas into a liquid that yields 'clean' gasoline, diesel fuel, or any of the slate of refinery products derived from crude oil..

It is frequently stated that the Norwegian Troll Field will be in "full" operation through the first quarter of the 21st century, however some observers are prone to claim that it might be delivering substantial quantities of gas in 80 years. This suggests a production profile of the kind shown in Figure 4.3, which also indicates the cost outline associated with producing gas. (This diagram also applies to the production of oil). What we have are investment costs to begin with, while later the cost structure is mostly dominated by variable (e.g. labor) costs and maintenance costs.

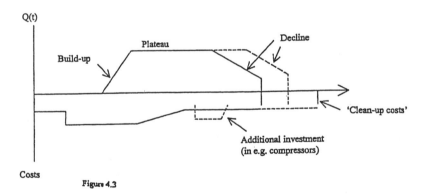

Figure 4.3

Some comments are necessary here. After the deposit has been 'exhausted', or nearly exhausted, a measure of 'clean-up' work will usually be necessary. Some environmentalists are beginning to say that the all-inclusive clean-up costs for very large offshore installations are so large that the firms operating these installations will do everything possible to avoid paying them, regardless of the environmental damage that may result. (In the case of nuclear energy, as you may remember, the clean-up process may turn out to be so expensive that, in some countries, nuclear electricity is actually very expensive electricity.).

The thing that everyone looks forward to, of course, is a long period of plateau production, when the gas is going out, and the money is rolling in. During most of that period, the majority of costs will be variable (e.g. labor) costs, to include maintenance costs associated with the depreciation of equipment.

Gas fields are sometimes described as having a e.g. 15-year depletion period if the annual output on the plateau is one-fifteenth of total recoverable reserves. This plateau output level might be held for considerably less than 15 years, although gas might flow for more than 15 years, taking into account both the build-up and tail-off (i.e. decline) periods. The advantage in looking at the problem in this manner is that it allows attention to be focused on two important variables: the height of the plateau (i.e. plateau output), and its length. The same is true for oil.

The amount of gas that is regarded as recoverable is that which can be directed from a deposit into pipelines. Most of this flow is made possible by pressure differences between the gas reservoir (or trap, as it is sometimes called) and the pipeline entrance that results from reservoir pressures of up to 10,000 pounds per square inch ($\#/in^2$). A gas reservoir can often be emptied of more than 80% of its contents (as compared to an average of 32% for oil), with as much as 50% of this attributable to gas expansion alone. For instance, Algerian gas once came out of its wells at such high pressures that no compressor stations were needed to move the gas the more than 400 miles to the Tunisian border. The rule is that delivery can take place as long as a suitable pressure differential is maintained between the pipeline entrance and the point of delivery. (This should be kept in mind when considering the discussion below about pipelines). Generally, pipeline networks consists of high pressure transmission pipelines and low pressure distribution systems, with the latter buying from the former.

If we consider on-shore gas fields (where a field can be taken as consisting of a number of reservoirs, while all the fields within a 1,000 to 10,000 square mile area comprise a basin), the first step in establishing a system which can deliver large amounts of gas over an extended period is to drill a number of exploratory wells over the estimated area of the field, so that an estimate can be obtained of the recoverable reserves of the field. With this estimate available, the gas in the field can be sold (usually on very long term contracts) and drilling competed. Should it be necessary, compression equipment can be installed to boost pressures in the vicinity of the wells. The number of wells drilled and the drilling program depends upon the desired production profile, which in turn is a function of the agreement between owners of the field and the buyer(s) of the gas, and almost always involves a uniform or increasing rate of delivery over a comparatively long period before the decline begins.

A scheduled level of sales by producers is a datum, and not always a fact. In 1972, in the US, an unforseen decline in production from existing fields led to industrial buyers temporarily laying off very many employees due to the diversion of their usual supply of gas to top priority residential users; but even so many communities were on the fringe of 'gas-outs' that would have left tens of thousands of homes cold. Later there was a period when large natural gas consumers – e.g. distributors and industries – refused to take gas that was contracted for, despite take-or-pay contracts. What happened at that time was that these gas buyers were sued for breach of contract, but they were able to argue successfully in court that they should be free to buy cheaper gas – gas that, in reality, only became available because some gas sellers made very expensive investments on which they were deprived of their expected yield by ad-hoc legal decisions.

Some observers claim that the greatest benefactors from all this were the lawyers who found themselves handling billion of dollars in take-or-pay litigation. In any event, it was about this time that natural gas deregulation in the United States picked up steam, and it eventually spread to Europe where the deregulation that has been proposed is so complex that it will have to be policed by other regulations that are, apparently, equally as complex.

Of particular importance in any discussion of natural gas supply in any mature gas province is the declining productivity of new gas wells, which means a larger number of wells and sharply increasing expenditure are required to obtain equal increments of output from the exploitation of existing or new gas fields. This is not just due to such things as well-clogging, falling inter-well pressure differentials, and the increasing density of wells in a particular gas field (which means that with a fixed amount of gas in a gas field, new wells will tend to feature a decrease in output), but the actual scarcity of gas down to the maximum depth at which gas is currently found in most of the known gas basins.

In the United States the number of wells drilled increased from 9,850 in 1977 to more than 16,000 in 1982, and during the same period the average well productivity of 115 Mcf/y may have been reduced by almost one-half. This situation implies that maintaining a constant level of production will involve finding and bringing into production an increasing number of wells. In the long run, obviously, this is asking for too much, and it is one of the reasons why the lowest growth rate in the gas producing world is to be found in North America, where production in 1994 was only up by 3.4 Tcf (or16%) from the 1984 level. Most of this growth, incidentally, was in Canada, which steadily increased its exports to the United States in the last decade of the 20th century, and has high hopes for an even more favorable sales climate during the 21st century. To see exactly how serious the situation is in the United States, readers should be aware that gas has become the primary target for drilling in the US., with more than 50% of all rigs committed to gas wells. In 1996 the reserve-production ratio has almost fallen to 7 years. With all due respects, the energy situation in that country should not give rise to an aggressively lighthearted attitude on the part of its citizens, nor should they place an overwhelming faith in such things as derivatives markets (e.g. futures and options) to secure for them the energy resources that are essential for both economic and social progress.

That brings us to the next step in the supply chain: pipelines. Perhaps the best way to begin here is to ask what is the ideal situation for bringing gas from producers to consumers? The simplest answer is to have huge gas deposits that are connected to the consuming area by pipelines with a very large diameter. Large pipelines allow the maximum exploitation of economies of scale in transmission; while large deposits mean that the pipeline and its compression equipment does not have to be abandoned early in their 'lifetime'.

Figure 4.4 shows a portion of a pipeline. From an engineering point of view its construction is a fairly routine operation; but beginning in the 1980s the politics and economics of such systems became greatly complicated. First there was the problem that President Ronald Reagan had with various European firms and governments over the amount of gas that could be purchased from the FSU, which on the technical level reduced to a dispute over the size of pipelines and compressor equipment, as well as who would provide the equipment; while during the nineties the deregulation debate moved into the spotlight.

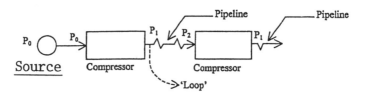

Figure 4.4

In this figure the pressure at the source is P_0. The compressor raises the pressure to P_1, but due to such things as friction losses in the pipeline, the pressure eventually falls to P_2, and another compressor raises it again to, e.g., P_1. As an example of this sort of thing, I can cite the pipeline from Australia's North West Shelf to Perth, the capital of Western Australia. This pipeline is 26 inches in diameter and runs 940 miles. There are five compressor stations through which the pipeline runs, as well as about 590 mainline valves. Both the compressor stations and the valves are computer controlled from Perth. As for the 'loop' in the diagram, this is a 'parallel' section of pipe that is sometimes added in order to increase capacity. In some situations it can be preferable to increasing the size of the pipe.

China has large supplies of gas, and the pipeline system is being rapidly expanded. The first large diameter pipeline was built in Sichuan Province in 1963, and at the present time there are about 12,000 kilometers of gas transmission lines of various sizes in service in the country. (Note: a 'transmission line' should be distinguished from a 'distribution line', where the latter usually carries gas – or electricity – to final consumers.) At the present time about 43% of China's gas is used for fertilizer production, 32% is used in the refinery sector or consumed in one way or another in the gas and oil sector, 14% serves as fuel in the industrial sector, 8.5% is consumed in the urban sector – mostly by households and small businesses – and the rest goes to minor uses.

In formulating a traditional production function for the above system, the key inputs (in *our* analysis) will turn out to be line pressure, which is a function of the size of the compressors, and the diameter of the pipeline. As it happens, scale economies are possible with both the compressors and the pipeline. With, for example, a given line pressure, increasing the amount of gas to be carried will call for increasing the diameter of the pipe; however the increase in the cost of materials for a larger diameter line will increase the quantity of gas carried by such an amount that the unit cost of gas transmission falls.

The reason for this is that the carrying capacity of a gas pipeline is proportional to the interior cross sectional area ($= \pi r^2$). On the other hand, the amount of pipeline material (e.g. steel) used is proportional to the circumference of the pipe (which is $2\pi r$). If the input of steel is doubled, then carrying capacity is more than doubled – at least up to a point. Compressor equipment also displays increasing returns to scale, though not indefinitely. With this equipment we can expect the U shaped cost curves – or distorted U shaped curves – featured in our elementary textbooks to apply.

It has recently become fashionable to claim that returns to scale in natural gas systems are overrated or insignificant, and that even the fragmenting of existing systems, to include the adoption of smaller pipelines and compressor stations, will not reflect unfavorably on costs. This is a myth, and reflects a basic contempt for the laws of both physics and economics.

Now for a few technical matters. As you found out in the first course in economics, a production function relates inputs to outputs: put simply, it takes inputs to produce outputs. Usually the output is labeled Q, and the inputs K and L, which stands for capital and labor. Symbolically we have Q = G (K, L). Something that needs to be emphasized here is that these production functions are conventionally *value added* production functions, with the arguments (on the right hand side) of this relationship limited to capital, labor and land – although land is usually assumed away. Intermediate products such as the output of other firms in the economy, to include natural resources, are not included explicitly among these arguments – although, of course, the output of the firm (Q) can be a natural resource such as crude oil, coal, etc.

One of the implications of the above discussion is that a graphical representation of a production function requires more than two dimensions, but we get around this by employing what mathematicians call level curves, and economists call isoquants. We take a given value of Q, e.g. Q_1, and show all the values of K and L that generate Q_1. This arrangement can be seen in Figure 4.5a, where $Q_1 = G(K_{11}, L_{11}) = G(K_{21}, L_{21}) = \ldots$

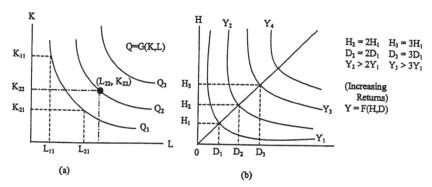

Figure 4.5

We want to do the same thing for natural gas. In your basic course in economics, the author of the textbook you used would have had no qualms about using K and L as inputs, but unfortunately we have to be more sophisticated. Being more sophisticated in this case means searching the advanced literature until we find something that can be simplified to fit into our discussion.

The 'something' we need is provided by Hollis Chenery (1949, 1952). His system consists of a gas source together with the pipeline entrance, and further down the line compressors and more pipe. The system presumably terminates at the beginning of the distribution phase. Where the flow, Y, from a given length of pipeline is concerned, we can write:

$$Y = F_1(D, P_1, P_2) \tag{4.1}$$

D is the inside pipe diameter, while P_1 is the inlet pressure to the pipe just after compression, and P_2 is the outlet pressure, as shown in Figure 4.4. Chenery also formulates an expression for compression, which is:

$$H = F_2(P_1, P_0) \tag{4.2}$$

P_0 is the inlet pressure to the *first* compressor, while H is the (e.g. horsepower) rating of the compressor. Obviously, *ceteris paribus*, it takes a certain horsepower (i.e. 'work') to raise P_0 to P_1. Next, in his interpretation of Chenery's work, Smith (1961) says that it is assumed that $P_0 = P_2$: the inlet pressure to the first compressor is equal to the outlet pressure of the pipe before the gas enters the next compressor (or, perhaps, before it enters the low pressure distribution system).

This seems perfectly straightforward until we examine figure 4.4. Then we see that there is no 'engineering' reason for P_0 to be equal to P_2. Instead it might have been better to forget P_0 for the time being, and to assume that for the purpose of this exercise the pipeline was configured in such a way that logically $H = F_2(P_1, P_2)$, and this applies for all compressor stations except the first: the first station raised P_0 to P_1. Why wasn't this done? The answer is that by assuming $P_0 = P_2$, we greatly simplify our work, to put it mildly.

Chenery also presents an "auxiliary" relationship between P_1 and the thickness of the pipe that can be ignored for the time being. Now let us look at our equation system, making sure that we understand that our primary goal here is a mainstream textbook exercise on the level of intermediate microeconomic theory, which captures at least some of the features of real world pipeline systems. Needless to say, the approach being taken in the present analysis applies to oil as well as gas.

$$Y = F_1(D, P_1, P_2) \tag{4.3}$$

$$H = F_2(P_1, P_0) \tag{4.4}$$

$$P_2 = P_0 \tag{4.5}$$

$$P_0 = \bar{P}_0 \quad \text{(A geological constant)} \tag{4.6}$$

Using (4.3) and (4.4), we can transform (4.1) and (4.2) to:

$$Y = F_1(D, P_1, \bar{P}_0) = F_3(D, P_1) \tag{4.7}$$

$$H = F_2(P_1, \bar{P}_0) \quad \Rightarrow \quad P_1 = F_4(H) \tag{4.8}$$

Note what has been done above. $\overline{P}_0$ was substituted for P_2 in (4.1). Since $\overline{P}_0$ is a constant, it does not have to be put in the kind of *implicit* relationships that we are dealing with here. Next, (4.2) was solved for a unique value of P_1, assuming that in the present case this operation was possible, which seems quite reasonable; and since once again P_0 is a constant, it can be dispensed with: its influence on P_1 is captured in the function, F_4. Now, P_1 from (4.6) can be put into (4.5), and we get:

$$Y = F_3(D,P_1) = F_3(D,F_4(H)) = F(D,H) \qquad (4.9)$$

Equation (4.7) 'looks' right, both in the engineering and economic sense. At the same time, however, I have certain reservations about this procedure. After examining Figure 4.4, I would have been tempted to begin the analysis with $P_1 = G_1(H,P_2)$, and $Y = G_2(D,P_2)$, and then tried to obtain equation (4.7). However it must be admitted that the above scheme conforms to most engineering relationships governing the flow of gas in a pipe. (See C.H. Paulette in *The Pipeline Engineer*, March 1968.)

Notice too how the function designations, the F's, are changed as we make our substitutions. Equation (4.7) is a production function of the familiar type that you should remember from your basic course, and corresponds to the one in Figure 4.5. At this point, however, a warning seems justified. Chenery's well known articles were brilliant examples of applied microeconomics, as is Smith's book, and we have every right to be impressed with their work; but not so impressed that we confuse it with engineering. What they end up by doing is proposing a relationship between inputs and outputs (i.e. a production function) that, perhaps, could be the object of some empirical/experimental work in economics or engineering. Chenery, for example, turns the implicit expression $Y = F(D,H)$ – or, if you prefer, $Y - F(D, H) = 0$, or even $F(Y, D, H) = 0$ – into a rather awkward explicit expression:

$$F(Y,H,D) = \frac{D}{k_1} + \frac{k_2}{k_1}Y - K_1 H^{5/3}\left[\left(\frac{D}{k_1 Y} + \frac{k_2}{K_2}\right)^2 - 1\right]^{1/2} = 0 \quad (4.10)$$

The k's in (4.8) are constants, as are the K's, although in Chenery's work the thickness of pipe (which is given) is in the equation, while here it is incorporated into K_1. What we can do now is to take some different values for Y, and plot values of H and D which will yield that Y. This rather onerous assignment will give us an isoquant system of the type shown in Figure 4.5b.

The system, as noted, is characterized by increasing returns to scale: if we double inputs, we more than double outputs; if we triple inputs, we more than triple outputs, etc. The same thing is true of an isoquant system constructed from equation (4.8), although it would not be easy to prove.

Getting increasing returns in a small pipeline involves no more than increasing the diameter of the pipe, at least to begin with.. Getting increasing returns in a pipeline *system* of perhaps several thousand kilometers, where friction losses can be considerable, and expensive compressors have to be used to maintain pressures, is quite another matter. The consensus seems to be, however, that if there is a very large gas deposit, and a very large customer base, the larger the diameter of the pipe the better – up to a point. That point seems to be about 65 inches (=165 cm) when this was written, although it has been said that new materials and technologies can raise that figure. This might also be an appropriate time to remind the reader that 'increasing returns' means increasing returns to scale or, what is the same thing, decreasing unit costs. The compressors could be expected to have the kind of U shaped cost curves that you encountered in your basic course in economics, and thus when they are a component of very large diameter pipelines, it is possible that they are operating on the 'upward portion' of their cost curves, where there might be sharply increasing costs. What this means for the entire system is not easy to say.

In the above discussion it seemed useful to dispense with one of Chenery's equations: $P_1 = 2ST/D$. What this expression does is to give us the maximum pressure that we can have in a pipe of thickness T and diameter D. This is important because the rule is to transport an equivalent amount of energy, a pipeline for gas should be both larger (in diameter) and stronger than one for oil or petroleum (i.e. refined) products. The reason is that even when under considerable pressure, natural gas contains less energy that the same volume of crude oil, and so high flow rates are essential. But high flow rates imply both high pressures and large diameters, which in the light of the above equation means a large T, and thus very heavy pipes. Once again we find ourselves in a situation where the cost disadvantages of extremely heavy/thick pipes might outweigh the advantages associated with a very large diameter, and help to account for the termination of increasing returns.

One thing remains here, and this has to do with the 'loop' shown in Figure 4.4. What this is all about is expanding an existing system. The compressors can be supercharged, and parallel sections added to the existing line. It has been suggested that under certain circumstances, an expansion of this type can result in a loss of economies of scale, depending on the length of the loops, the additional compressor investment, and the increased amount of gas that is expended in conjunction with the increased compression, since gas from the pipeline provides the energy input used to drive the compressors. Unfortunately, this matter can only be settled on a case-by-case basis, since in many situations looping has a great deal to offer.

This section will be concluded by saying something about storage – to be exact, the bulk storage that takes place outside of pipelines.

Occasionally storage will play a prominent part in discussions concerning *firm* and *interruptible* service, especially where industrial sales are concerned. Firm gas is supplied throughout the year at a more-or-less constant flow rate, while interruptible gas is primarily available during *off-peak* periods (when there is excess capacity). Obviously, if there is gas in storage, then it can be taken out to satisfy 'extra' demand at almost any time, and thus storage is (within limits) a substitute for production capacity.

The expression 'off-peak' was used above, and so this might be the time to distinguish between two types of peak-load situations: *firm peak* and *shifting peak*, both of which apply to gas (and electricity) systems where full capacity is not utilized during some periods of operation. A firm peak exists when the peak (load) does not change as a result of changing the price of off-peak gas to any level: off peak gas in these circumstances is not a substitute for firm peak gas. By way of contrast, a shifting peak means that a lowering of just variable costs to off-peak customers will cause them sharply decrease the amount of gas demanded during peak periods. The load pattern is altered, and perhaps to a considerable extent. Another important term is *peak shaving*. This can involve the injection of supplemental supplies of natural gas from underground storage (or even LNG facilities) into a pipeline system in such a way as to avoid expanding the system (via capital investment) when the potential maximum demand appears. Put another way, the peak might remain, but it would be 'flattened' and perhaps extended over adjoining periods, and it would not become more pronounced..

The logic behind storing gas is simple. Demand is not constant over the year, and usually rises during the winter. Having gas available to meet this demand can mean substantial profits for the owner of the storage facility (as mentioned in the comments above on US-Canada pipeline trade). Of course, storage is not costless, and given the amounts of gas usually involved, there is ample opportunity to make catastrophic errors of judgement.

In 1995 Exxon spent 600 million dollars to develop 3 storage facilities in Holland and Germany, which indicated that storage had become a very important commercial strategy. Exxon's facilities provide 120 Gcf of storage and 6 Gcf/d of withdrawal capacity.

These installations, along with others being constructed, are intended to make Holland *the swing supplier* for Europe: they will make gas available whenever its price exceeds a certain level, just as the United States is often regarded as the swing producer in the world coal market, and Saudi Arabia in the world oil economy. Having these facilities in Holland might also allow postponing the large costs associated with installing compression at Groningen as gas (well) pressure declines, and thus output falls.

Partially depleted gas fields are capable of being used as storage sites, but they can take a long time to fill and empty, and thus lack flexibility. Salt-cavern storage sites that can be filled and emptied in less than a month are generally more efficient, but they require a larger monetary investment. Since Exxon and a few other large firms are ostensibly jockeying for position in the gas-on-gas competition that the Energy Directorate of the European Union envisions as a prime building block of Europe's energy future, their strategy is to have as many storage sites as possible over as large an area as possible. This is almost certainly a good business move in the short run, although what it will turn out to be in the long run when Norwegian, U.K., and Dutch supplies are palpably incapable of supplying European demand, and firms like Exxon must rely on Russians gas and LNG to service these storage sites, remains to be seen.

Storage often appears to have broken the ability of violent changes in the weather to alter gas prices by a large amount, since the ability of demand to surge beyond a gas system's capacity has been greatly limited. But at the same time there will almost always be regional price 'spikes' because of the difficulty of delivering gas. It is this difficulty, incidentally, that provides one of the main stumbling blocks to turning gas markets into 'textbook' competitive markets.

One of the most stimulating topics in energy economics is short term oil prices, and the same might be true with gas if gas spot and derivatives markets functioned the way that certain influential persons believe they function. This topic ties in with storage via the concept of natural gas *market centers* where, as with 7-11 type stores, prices would be visible for spot gas, transportation services at various times, storage, and relevant derivatives. In fact, in ideal circumstances, storage sites would be an integral part of market centers while, physically, the centers themselves would take on the appearance of *hubs* where a number of pipeline converged, and in addition where gas not being moved between buyer and seller could be 'parked'.

Storage improves the efficiency of gas systems and, *ceteris paribus*, lowers the cost to consumers in the aggregate. It remains an open question, however, whether the presence of even adequate storage can be used as one of the justifications for reorganising gas markets in the manner being proposed by the European Union's energy directorate.

Before moving to another facet of this topic, it might be useful to look at the Latin American gas scene, which is a key part of the integrated energy market that is taking shape in that part of the world. The most ambitious project thus far appears to be the Bolivian-Brazil pipeline, whose shareholders include Enron and Shell. This pipeline cost slightly over 2 billion dollars, and is 3,200 kilometers long: from Santa Cruz in Bolivia, east to Sao Paulo in Brazil, and then back to the important Brazilian city of Porto Alegre. Notice these figures: they amount to $625,000/kilometer.

Ostensibly, this pipeline would lead to the construction of power plants and industries throughout the region (why?). Another project often mooted involves the Camisea gas field, which is the largest gas field in Latin America. Fifteen years have passed, however, without a major exploitation of the field, because local demand does not justify building a pipeline as costly as the one mentioned above – nor has it been judged that a smaller one would make sense if any genuine development is going to take place within three or four thousand kilometers of the field. The ideal export market is thought to be Brazil, where the energy shortage has been described as "chronic".

In contrast to e.g. Europe, where major investments in gas are scheduled regardless of the regulatory climate, deregulation and/or privatisation are important in Latin America in order to attract the billions of dollars in foreign investments that are apparently essential in order to optimise the exploitation of energy resources. In Argentina and Chile, market liberalisation has been underway for a decade, but whether it will take hold in the countries to the North remains to be seen.

4.2.1 Exercises

1. The price of crude oil during 1990-95 averaged about $18/b. How much was that in $/MBtu?
2. Why would LNG become so popular when it costs more than pipeline gas? The pipeline from Bolivia to Brazil mentioned in the last part of this section costs a great deal of money. Why bother with this pipeline – why not just build a large nuclear plant in Brazil? Gas from Argentina will go to Uraguay, where it will be used to generate electricity, and the electricity will be transmitted to Brazil. What are some of the things that could explain this scheme for supplying Brazil with electricity? Why not just build a pipeline from Argentina to Brazil?
3. It has been stated that many producers of natural gas – and in particular gas located in arctic waters and other remote locations – are unwilling to trust their fortunes and those of their share and bond holders to the spot market. What spot market? What do they prefer, and why? Discuss!
4. Suppose that we have $Y = H^{2/3}D^{5/3}$. Do we have increasing, decreasing or constant returns to scale?
5. Suppose that we had the following equations:

$$Y = a_1D + a_2P_1 + a_3P_2$$

$$H = b_1 P_1 + b_2 P_0$$
$$P_2 = P_0$$
$$P_0 = \overline{P}_0$$

In case readers have forgotten, the last expression means that P_0 is *exogenous*: it is given, a *market circumstance*, where the word itself has Greeks roots which mean "born outside of". With this settled the reader can solve for Y as a function of D and H! Now, assume that all the a's and b's are positive. Are you comfortable with this assumption? Give the a's and b's some numerical values, and draw 3 isoquants! Can you say anything about returns to scale here?

4.3   Regulation and Deregulation

The arguments in favor of deregulation are clear, but so are the arguments against. The principal argument for deregulation is that the way things are going in the world today, markets appear to be right more often than regulations and regulators. This is probably true on many occasions, but it is certainly not always true. The basic problem here is that many real life markets not only *do not* function with the flexibility and efficiency that they display in our textbooks, but they cannot; and when these situations arise, the regulators must be called in.

Sir Alan Walters – formerly personal advisor to the very conservative Prime Minister of the U.K., Margaret Thatcher – lists some of these situations in his celebrated textbook (co-authored with Richard Layard). He says that government intervention is "normally suggested" when there are increasing returns to scale, indivisibilities, technological external effects, and/or market failure connected with uncertainty.

The key word above is 'normally'. What it means is that there can be situations in which regulation is inadvisable, even though all the above named irritations are present to some extent or another. The problem is detecting, acknowledging, and/or estimating their strength and scope. This is one of the reasons why John Maynard Keynes said that "economics is an easy subject that is difficult".

The principal issue with natural gas deregulation – and to a somewhat less degree electricity – is making sure that specialized, irreversible investments can be efficiently used. The way this had been done in the past is to enter into long-term contracts which remove the uncertainty associated with these investments. Among the things this meant for natural gas was that pipelines of an optimal size could be constructed because buyers had been secured 'up-front' for the full capacity of these conduits. For reasons that should be clear by now, this arrangement has resulted in a minimizing of production and transportation costs, which has helped to reduce the price of gas to final consumers. Natural gas has almost always, everywhere in the world, cost less that the thermally equivalent oil price.

Bill Tilden, the great American tennis player, offered his admirers and students the following rule:"always change a losing game; never change a winning game", and in those countries and regions where comprehensive regulation was a losing game it should have been changed; but as bad luck has it, attempts are being made to change it everywhere. For example, the firms that have supplied most of Western Europe with plentiful supplies of low-cost gas for many years are to be 'fragmented'. (The same is true in many cases for electricity: in Sweden and Norway, which have the lowest electricity costs in the world, deregulation – i.e. 'changing the game' – could eventually mean rising electricity prices for all final consumers of electricity except the largest firms. These firms have considerable financial muscle that they can use in their negotiations with power suppliers. But as noted elsewhere in this book, the uncertainty caused by deemphasizing long term contractual arrangements could also work against the largest buyers, especially if sellers have considerable market power.)

The most incisive academic observer of natural gas deregulation is probably Professor David Teece of the University of California. In a seminal paper (1990), Teece traces the history of deregulation in the United States. To his way of thinking, the opening stage alone of comprehensive deregulation has "jeopardized long-term security and created certain inefficiencies." Teece also denies that a series of short-term contracts (and/or spot transactions) can substitute for vertical integration.

Most important, he identifies what he considered to be the biggest deregulatory blunder of all: *open access* – which is sometimes confused with *common carriage*. Common carriage involves the obligation of a carrier (e.g. a pipeline or an electric power line) to provide access to the capacity managed by the carrier on a pro-rata basis. If, for example, a new transactor requests carriage, he or she must be accommodated, even if it means reducing the carriage provided other transactors. By way of contrast, open access (contract) carriers, such as most gas pipelines, provide access on a first-come first-serve basis to all clients willing to pay the pipeline's maximum tariff. Once capacity is fully utilized, the carrier must refuse new clients.

Before we plunge even deeper into this subject, with new definitions and names, readers should think about the following: unfavorable outcomes *ex-post* for a transactor do not imply *ex-ante* inefficiency on the part of regulators and/or market actors. What it can mean is a change in the economic environment that was unforeseeable by regulators and market participants alike, and which, when experienced, could not be accommodated in the short run because of the durable nature of the very expensive structures and machinery used in the natural gas industry.

Without knowing it, many deregulators and their supporters want gas production and transmission facilities to display the most attractive features of financial markets. These include the rapid dissemination of information; frictionless transfers of the relevant asset; large numbers of transactors (buyers, sellers, brokers, etc); together with dynamic contingency (i.e. derivative) markets which can mobilize the liquidity required for efficient hedging (and speculation) to take place.

The presence of all or most of these characteristics might indeed make the proposed deregulation (i.e. reregulation) an attractive proposition, but given the constraints in actual gas markets, some doubt exists as to whether this ideal could be realized. Consider, for example, the following oversimplified presentation of one of the most annoying problems that the proposed changes could evoke.

Someone comes to your door to sell you a bundle of natural gas, which he may or may not own at the present time. But whether he owns it or not, he still has to obtain the pipeline space to get it to you. Or, working from another direction, Ms Saleswoman obtains the pipeline capacity needed to transport 'X' units of gas, but when she knocks at your door, you tell her in a friendly manner that you have all the gas you need. How are 'matching problems' like this to be solved? One way is to have a central computerized auctioneer that simultaneously determines prices for gas at all locations, as well as complete price information for transportation services. In these circumstances, interested parties could sell and buy gas and pipeline space on the basis of prices appearing on their computer screens. In a fully developed system, ideally, they might be able to obtain a variety of forward, futures and option prices.

Teece, however, is able to place some of the aspects of this rosy vision into relief when he quotes a player in these markets who says that arranging transportation in the United States interstate gas industry involves "chaotic monthly nominations to pipeline transport controllers". Furthermore, with major pipelines now having almost 1000 customers, where before they had 20 to 30, the cost of administering the system has increased by up to 15%. More important, despite superhuman around the clock efforts, it is an "anomaly" if buy and sell orders are matched. Accordingly, huge imbalances are commonplace.

Commonplace, and to a considerable extent incurable. The inescapable difficulty is that natural gas is not always *fungible*. Fungibility means that certain assets can always or almost always be substituted for each other – e.g. the currencies of Western European countries tend to be fungible because of such things as high speed transfers, and arbitrage that keeps prices in line. (Arbitrage has to do with the law of one price: in a single market there is only one price for a good or service. This is usually managed by buying cheap or selling 'dear'.) We do not have this happy situation with gas. The price of gas at A may not be the same as the price of a chemically identical gas at B, and the arbitrage that could equalize their prices – taking transportation prices into consideration – may not be able to take place due to the absence of a pipeline between A and B, or the absence of sufficient capacity in an existing pipeline.

Gas deregulation in the US was introduced and unconditionally supported by the same sort of individuals who have taken up the cudgels in Europe. For instance, when the U.S. Federal Energy Regulation Commission (FERC) claimed that it had conducted an economic analysis proving the virtues and efficiency of deregulation for both gas consumers and the national economy, it was immediately sued by an anti-deregulation lobby (Citizen Action), and forced to admit that in reality no analysis of a 'scientific' nature had taken place.

Nor is one likely to take place, given the high probability of the wrong outcome.

4.3.1 Exercise

1. Gas consumption in OECD Europe + (Czech Republic, Hungary, and Poland) is estimated to rise from 312 Gcm in 1992, to 423 Gcm in 1998, What is the average annual rate of increase over that period.

4.4 Marginal Costs and Peak Load Pricing

Peak load pricing is usually discussed in relation to electricity, and often employing the kind of 'single technique' diagram that will be used below. Some questions must be raised, however, as to how effective this approach is, since almost all large electricity generating systems are characterized by multiple technologies. The contention here is that depicting natural gas systems as single technique technologies can occasionally be a useful pedagogical departure.

The main discussion in this section features a model with a long history in the literature of economics. Ray Rees' celebrated book (1984) draws heavily on this model, and elements of the model are employed in perhaps the most elegant chapters of one of the most elegant microeconomics texts – that of Layard and Walters (1978). I have recently recently employed this kind of construction to discuss electricity pricing in Sweden, although in truth it is more suitable to the classroom than the real world; however it is valuable for introducing students (and others) to some complicated theoretical aspects of electricity pricing situations.

In Figure 4.6 there is a brief scrutiny of peak loading problem concerning a private enterprise – e.g. Luna Park in Sydney, or Riverview Park in Chicago; but most of the literature on this topic deals with public enterprise (i.e. public utilities): demand curves are not 'flat', or perfectly elastic, but take on the appearance of the demand curves in the chapter on monopolies in your 'fundamentals' textbook. There are, however, no marginal revenue curves. The reason for this is what Layard and Walters call "the pricing rule", by which they mean the pricing rule for public policy analysis: "Find the output which, for the existing plant, equates the demand price to the marginal cost, and then charge the marginal-cost (and demand price) for that output." When they say marginal cost in that passage they mean the short run marginal cost.

This sounds perfectly straightforward, particularly when we remember that both Richard Layard and Alan Walters are important economists. The problem is, however, that the engineer-economists of Electricité de France prefer to think in terms of the long-run marginal cost. This matter will receive some attention below, but the reader is asked to remember that the theoretical apparatus being employed does not tell the entire story.

Suppose we start our analytical work by looking at pricing in a private firm in which the demand for output varies sharply during a given time period. Take, for example, a 'normal' week at an amusement park such as Luna Park in Sydney. What we could have is the situation shown in Figure 4.6. The demand curve for the peak period is D_p, and for the off-peak period D_o. The corresponding prices are this p_p and p_o, which – together with outputs – are determined where MC = MR for the appropriate curves. These prices maximize private profits, which might turn out to be very large, depending on the average (i.e. unit) cost at which production takes place. The reader should, in fact, draw an average cost curve, and make sure that he or she can identify the profit and, if possible, the 'welfare' losses that are associated with this kind of exercise.

100

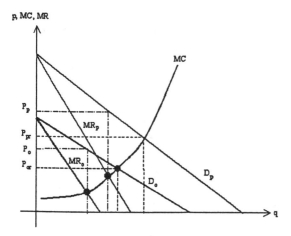

Figure 4.6

Now let us suppose that the Australian government decides to regulate amusement parks. They would then direct that the prices in the above diagram should be set at p_{pr} and p_{or}. This gives us marginal cost pricing, and to keep things simple we will not be concerned here with whether marginal cost means SMC or LMC (although if all 'adjustments' have taken place, it would be SMC).

Theoretically, this is the best arrangement for consumers, because it implies that now they are paying for a unit of output (entertainment services) the exact value of the resources used to produce the output. Naturally, we have overstepped the bounds of reality in this example, since we are implicitly assuming a completely homogenous output; however once again we can recall what one of the great mathematicians of the 20th century, Bertrand Russell, said about approximations: "Although this may seem a paradox, all science is dominated by the idea of approximation.".

As for producers/owners, if costs are taken into consideration the value of the best alternative use of the resources employed in the production of 'q' (i.e. opportunity cost), then the prices p_{pr} and p_{or} will contain a normal profit because, as should always be remembered, in economics (though not in accounting) *normal profit* is a cost. (If producers/owners cannot make a normal profit with a given enterprise, then in theory, they always have the 'opportunity' to abandon that activity and transfer the resources that they have tied up in it to 'safe' financial assets, such as government bonds. There are, of course, some problems here with irreversibility, however this aggravation should be taken into consideration before embarking on the particular project.)

Now let us go to the usual single technique representation, featuring 'perfectly complementary' or input-output type production functions. In Figure 4.7a and 4.7b we have the technique in question, with q = F (K, L) = min (K/a, L/b) as the equation for the production function. First of all, let us look at positions such as x and y, where we have no redundancy of capital or labor. At those positions we have K/a = L/b. In Figure 4.7b, for example, taking a=1 and b=2, and with K= 10 = $\bar{K}$, and L =20 = $\bar{L}$ we have q = min (10/1, 20/2) = min (10,10) = 10. 'Min', of course, stands for minimum, and the dimensions of 'a' are capital per unit of output, and for 'b' labor/output. It can also be noted that 'q' is in units per time period. It is a flow.

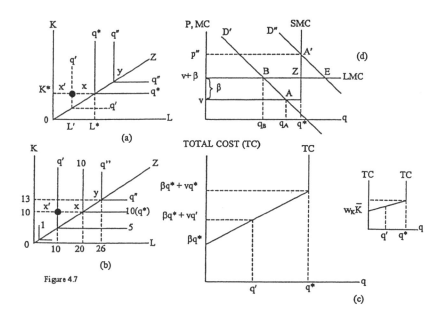

Figure 4.7

We continue by examining the situation at x', where K=L=10. From our production function we have q = F (K, L) = min (K/a, L/b) = min (10/1,10/2) =5, and at that location we can speak of a 'redundancy' of capital or, as it is sometimes put, 'an excess supply of capital'. (If K/a > L/b, then q = L/b).

As for costs, suppose that a unit of capital costs w_K dollars per time-period, and the cost of labor is w_L dollars/time-period. The time period that we usually employ in microeconomics is the year, but following e.g. Rees (1984), we use the day, which for this particular type of problem turns out to simplify a great many things. Continuing, if we assume a situation such as the one at x, but with q=1, we would need K=a, and this would cost $aw_K = \beta$. Similarly we would need L=b, which would cost $bw_L = v$. The dimensions of β and v are dollars per unit of output per time-period. (Exercise: q=1 is shown in Figure 4.7b. You can show a and b!)

The total cost of any quantity of output that is produced with an amount of capital K, and the amount of labor L, is $w_K K + w_L L$. If, for instance, we have $K = K^*$ and $L = L^*$, we then have (obtaining values of the w's from just above):

$$TC^* = w_K K^* + w_L L^* = (\beta/a)K^* + (v/b)L^* = \beta q^* + vq^* = (\beta + v)q^* \quad (4.11)$$

For any q on the ray OZ we have as the cost of q: $C = (v + \beta)q$. If this is clear, we can turn to the situation at a point such as x'. Here we have:

$$TC' = w_K K' + w_L L' = (\beta/a)K^* + (v/b)L' = \beta q^* + vq' \quad (4.12)$$

Once again the reader should pause and make sure that he or she understands this algebra, and in the light of Figure 4.7, exactly what it means. Notice in (4.10) that output is q' and not q*, and as a result we have $0 < q' < q^*$. One way of getting the total cost curve shown in Figure 4.7c is to start at K* in Figure 4.7a and proceed to the right toward x', and beyond to x. This increases the cost from $w_K K^*$ to $w_K K^* + w_L L^*$, or from βq^* to $(\beta q^* + vq^*)$, as output increases from zero to q*.

In examining Figure 4.7c, it should be clear that the short-run marginal cost is v, but if it is not clear, it might be instructive to derive it. Suppose that we have a given amount of output q, and it is possible to increase this by one unit ($\Delta q = 1$) by only increasing the variable factor, which here means labor. (Consider, for example, the situation at x'.) Our increase in cost, ΔC, is:

$$\Delta C = [v(q+1) + \beta q^*] - [vq + \beta q^*] = v = SMC = [TC(q + 1) - TC(q)] \quad (4.13)$$

To reiterate, we get the SMC because we only changed the variable factor, and we have $\Delta q = 1$. Thus, $\Delta C/\Delta q = SMC = v$. Or, the reader can work directly with equation (4.10). [But notice the following: in this derivation $v(q+1)$ means (q+1) *multiplied* by v; on the other hand $TC(q+1)$ means total cost as a *function* of (q+1)!]

Now let us get the long-run marginal cost by varying both factors in an optimal manner. This means starting at a point such as x or y: we would not start at e.g. x´, because here we can raise production by just increasing L, while if we just increased K we would not increase output, since we already have excess K. What about increasing both K and L at point x'. This would be meaningless, since we get the same increase in output that we would have obtained had we only increased L. Put in a more formal way, this was not an optimal move: it was wasteful, and therefore it has no place in mainstream economic theory – although wasteful gestures are not unknown in the real world. Taking the initial production as q*, and increasing it by one unit, we get:

$$\Delta C=(v+\beta)(q*+1)-(v+\beta)q* = (v + \beta) =LMC \qquad (4.14)$$

These marginal costs are shown in Figure 4.7d, where two demand curves, D' and D", have been added for the purpose of discussing the following situation.

1. The usual textbook version of long versus short run marginal cost pricing: i.e. in Figure 4.7d, pricing at B or A on demand curve D'.
2. An unusual but non-controversial case of short-marginal cost pricing at point A' on demand curve D".
3. Peak load pricing, where to keep the analysis simple we have two demand curves, e.g. D" (peak) and D' (off-peak), with each valid for 12 hours of the day. However, this situation requires a somewhat closer scrutiny than possible in Figure 4.7d. Certain things in that diagram are irrelevant, and merely get in the way; while others are extremely important and should be magnified.

As the reader will soon find out, from the algebraic and/or graphical point of view we have a perfectly straightforward story. Things become complicated when we must resort to words. SMC pricing (in this analysis) means producing at A, with price v, and output q_A. This is not particularly attractive from a business point of view, since we do not recover any of our capital costs ($\beta q*$). The way this shortcoming is usually approached is that anyone wanting access to q has to pay an *entry charge*. In addition they pay v for every unit they consume.

What we have here is an example of two-part pricing. How much should the (total) entry charges be? Well, what about βq_A? If this were imposed, then we obviously would have average cost pricing since $(\beta q_A+vq_A)/q_A = \beta+v = LMC = $ average cost, which is convenient. Please note, however, that we still have not recovered the total cost of the capital stock. We are 'deficient' by an amount $\beta q*-\beta q_A, = \beta(q*-q_A)$, which would very likely be added to βq_A in order to keep the shareholders from foaming at the mouth.

Regardless of what the entry charges are, we have departed from the kind of SMC pricing that purists like the late Professor Schweppes of MIT desired, where entry charges of all varieties and magnitudes were strictly taboo. For him, SMC pricing meant exactly that, and no less. On the other hand, Anna P. Della Valle once informed the present author that her version of SMC pricing included recognizing "constraints", by which she presumably meant granting the management of firms practicing SMC pricing the latitude to adjust their bills (when absolutely necessary) in the interests of avoiding insolvency. All this is well and good, but in some situations ploys of this nature bring the firm closer to average than to short run marginal cost pricing. Much closer.

A reference to an important contemporary debate might be relevant here. The popularity of SMC pricing among certain persons may be due to the fact that, in theory, this kind of pricing means that the price paid for all produced units of an item is the marginal cost of the *last* produced item. Suppose, therefore, that we have a supply curve – which the reader can draw – which indicates that a firm has the capacity to produce 100 units of an item for a unit marginal cost of $1, and 25 units for a marginal cost of $2. If demand is 110 units, then with SMC pricing 110 units will be sold for a unit price of $2, and total revenue will be $220. If average cost pricing were employed, total revenue would be (Fixed Cost+Total Variable Cost)/q). Suppose in this example that the fixed cost was fifty. Then the total 'allowable' revenue would be $100+20+50 = 170$, and the unit price that would be set if 110 units were to be produced is $170/110 = \$1.55$. How does this observation fit in with the previous discussion? In Figure 4.7, with SMC pricing and the demand curve D" the price would be p". With average cost pricing and the demand curve D" the price would be $v+\beta$. Which would you prefer? (Exercise: Identify the profit in both cases!).

The state of affairs at B in the same figure characterizes LMC pricing. In theory, consumers pay a per-unit price of $v+\beta$ for q_B, but once again the full cost of capital has not been recovered. Obtaining this will require entry charges of $\beta(q^*-q_B)$. Electricité de France (EdF), which ostensibly practices LMC pricing, might include all or a portion of this amount in the category 'financial charges'. One of their arguments for LMC pricing is that this extra charge is lower with (so-called) LMC pricing than with SMC pricing. The economists of EdF are the best in the world where these matters are concerned, but even so I find the results of their efforts unnervingly close to average cost pricing.

The second point above has to do with the intersection of the demand curve D" at point A', where SMC is greater than LMC, and the price charged consumers, p", is greater than $v + \beta$. There is no need to delve too deeply into this arrangement, although it implies to some observers the beauty of rationing a fixed output by means of the price system, since only those persons or enterprises who are 'financially deserving' will be able to enjoy access to the good, or at least as much of it as they desire.

But there is another valuable concept being dealt with here that will be of considerable use directly below. When $p>v+\beta$, as at A', there is a demand for *new* capacity: the value placed on new capacity is greater than its cost. The exact amount of new capacity desired is that which would provide an increase in output corresponding to point E (which the reader can show). There is no need, in an elementary textbook, to discuss at great length the massive problem of providing for the new capacity in a time of slow growth, and the sub-optimal solutions that this is liable to give rise to, but in the UK's newly deregulated electricity market, this topic cannot be avoided. Apparently, in order to encourage private firms to undertake capacity expansion when it is required, it has been decided that the 'equilibrium' price – whatever that means in this context – is to be the LMC price. This is indeed a surprise, because the gentleman in charge of deregulation/regulation was, during his academic career, a stout defender of SMC pricing. He even provided a celebrated mathematical proof of why it was the only correct mode of pricing the goods and/or services of utilities.

That brings us to the challenging next phase of our discussion of peak load pricing. To begin, let us repeat something that was said above in conjunction with Figure 4.7d: when $p > v + \beta$, it signifies that *new* capacity is warranted, but demand for some *existing* capacity is positive when $p > v$. The thing to fathom is that there is a valid argument for supplying new capacity only when the amount consumers are willing to pay for it is greater that its supply price (or $p - v > \beta$).

As an example of what we are talking about, consider the situation just to the left of q_A in Figure 4.7d, taking D' as the relevant demand curve. Here the value placed on capacity is positive (i.e. $p > v$) but not sufficiently positive for us to obtain $(p - v \geq \beta)$. By way of contrast, to the left of q_B, we see that $(p - v) > \beta$: the value placed on capacity is greater than its cost. Do we install new capacity? The answer in this case is *no*, because according to the diagram we already have the capacity available to produce another unit – the marginal unit – of the output q, and this is the object of the exercise. But at point A, with the relevant demand curve D", the value placed on capacity is greater than its cost, and we have come to the capacity limit, which corresponds to q*. Accordingly, capacity should be added.

In the exercise that we are now working toward, the assumption will be that we possess demand curves for q, and we want to make an optimal decision regarding capacity. What is an optimal decision in this context? If we forget about the uncomfortable fact that if we were dealing with electricity we always need to have some reserve capacity (since demand overloads can lead to blackouts), then in Figure 4.7d the optimal capacity would be associated with the point where the SMC and the LMC curves intersect at point Z. We also, of course, know the cost (per unit of output) of the variable factor, or factors, which is v, and so from here on it should be smooth sailing. (Observe: point Z should never receive any 'special' attention in discussions of this nature.)

Smooth, but not too smooth, because we no longer have one demand curve, but two: off-peak demand (D_o), and peak demand (D_p). It also needs to be appreciated that the demand curves we desire are not just demand curves for output, but also for the physical capacity to produce this output. To get these we save ourselves a great deal of trouble by assuming 12-hour periods for each category of demand. Thus, if the (marginal) cost of capacity in the total time interval (24 hours) that we are interested in is β, then the cost in each 12-hour period is $\beta/2$, and the optimal capacity condition can then be written as $(p_o - v) + (p_p - v) = \beta/2 + \beta/2$. This expression will emerge in the following discussion.

The order of attack here will be as follows. First an algebraic analysis which, hopefully, most readers will be able to follow, and whose ultimate aim is to show who pays for capacity, and how much they pay. Then there will be a (mostly) numerical example that goes over the same ground, and which should tie up all loose ends. Something that the reader should pay careful attention to in the following is the great advantage of being able to use linear demand curves: curves having the form $y = \alpha + \theta x$, where θ is the slope (or gradient), and is negative, while α is the intercept on the vertical (y) axis, and the intercept on the horizontal (x) axis is $(-\alpha/\theta)$. Some readers might like to take this opportunity to brush up their knowledge on this subject, in which case they should see what they can do about determining the slope and intercepts of the simple equation $p = 11 - (1/200)\,q$, and then make a rough drawing of it.. Other readers may not desire to have any part of this project, and I suggest that they go directly to the next set of exercises.

That brings us to Figure 4.8a and the demand curve D, which we obtain by first subtracting v from the off-peak and peak demand curves in order to get what a consumer is willing to pay for capacity q. Then add these curves vertically.

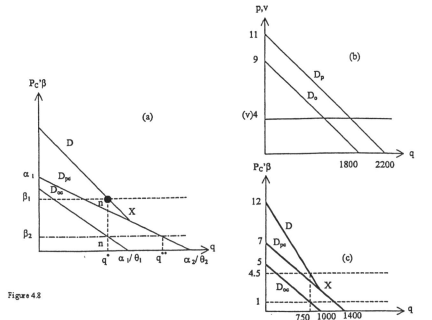

Figure 4.8

The algebra can commence by writing down the equations of the demand curves for q' (i.e. output, and not capacity q). These are shown in Figure 4.8b.

$$D_o: \quad p_o' = \alpha_1' - \theta_1 q_o' \quad \text{(12 hour, off-peak demand)} \qquad (4.15)$$

$$D_p: \quad p_p' = \alpha_2' - \theta_2 q_p' \quad \text{(12 hour, peak demand)} \qquad (4.16)$$

Next we go to the demand for capacity curves, which are shown in Figure 4.8a.These are obtained by subtracting v from (4.13) and (4.14)

$$D_{oc}: \quad p_{oc} = \alpha_1 - \theta_1 q_{oc} \qquad (\alpha_1 = \alpha_1' - v) \qquad (4.17)$$

$$D_{pc}: \quad p_{pc} = \alpha_2 - \theta_2 q_{pc} \qquad (\alpha_2 = \alpha_2' - v) \qquad (4.18)$$

These two expressions are added vertically in order to get the total demand for capacity, and then set equal to the 24 hour cost of the capital input.. In Figure 4.8a, notice the situation when the output is q^*, and how addition takes place: the interval n ($=D_{oc}(q^*)$) is added to $D_{pc}(q^*)$ to obtain $D(q^*)$. We now write:

$$(\alpha_1 - \theta_1 q_{oc}) + (\alpha_2 - \theta_2 q_{pc}) = \beta/2 + \beta/2 = \beta \qquad (4.19)$$

Since we have added vertically, we can set $q_{oc} = q_{pc} = q$. Thus we have for the left-hand stretch of D (i.e. to the left of X):

$$(\alpha_1 + \alpha_2) - (\theta_1 + \theta_2)q = \beta \qquad 0 \le q < (\alpha_1/\theta_1) \qquad (4.20)$$

The right hand stretch of D (which is the part of D_{pc} to the right of X) is:

$$\alpha_2 - \theta_2 q = \beta \qquad (\alpha_1/\theta_1) \le q \le (\alpha_2/\theta_2) \qquad (4.21)$$

From here we go to two possible values of β: β_1 and β_2. The first of these (β_1) intersects D in such a way as to give value of q equal to q* (q = q*). In these circumstances, both the peak and the off-peak consumers pay for capacity. We get the amounts they pay from the vertical distances up to their respective demand curves. This is the 'unit' charge, and the total charge is obtained by multiplying this by q*. To this must be added payments to the variable factor, which amount to vq*.

As for β_2, this intersects D to the right of X, where D coincides with D_{pc}. With this arrangement, only the peak demanders pay for capacity, although both groups pay operating costs of v for each unit of capacity they actually use.

A few comments on terminology and method are unfortunately necessary at this point. In the above exposition the words capacity and output have, to a certain extent, been used synonymously, although what is happening is that given demand and capacity cost, the capacity (q) is the object of the exercise. But we obtained the cost of capacity (β) from the cost (or rent) of capital (w_K), and so we can look at it another way. Given the cost of capital (and the cost of the variable factor), we choose a certain capital, and thus capacity (e.g. q*), which in turn permits a certain output. Why go about it this way, and not with isoquants and isocosts, as in the introductory course? The answer is that since we are concerned with peak load pricing here, it is convenient to minimize the role of production factors (i.e. inputs). This does not mean ignoring the fact that (linguistically) capacity is more intimately related to the capital input than to output, but that we need to determine who is to pay for the equipment that produces the output (e.g. gas). In the last sentence of the previous paragraph, for instance, what we are saying is that when output in Figure 4.8a is q**, capital is paid for by peak consumers, and where this particular output is concerned off-peak consumers get a free ride – although they must pay operating (variable) costs for the output they consume. (And don't forget, the discussion here concerns marginal cost pricing in an *ideal* setting. Thus, it only yields some guidelines to real world pricing; even though in certain situations these could be very valuable guidelines.)

The final step in our presentation is a numerical example. We will take v = 4, and β = 4.5. For the half-day demand curves take $p_p' = 11 - (1/200)q_p'$ for D_p, and $p_o' = 9 - (1/200)q_o'$ for D_o. By setting the p's equal to zero we get as the 'ranges' for the q's: $0 \leq q_p' \leq 2,200$, and $0 \leq q_o' \leq 1,800$. These values are shown in Figure 4.8b; they can be considered the demand for e.g. gas or electricity – i.e. *output*.

Next subtract v (=4) from these demand curves, and we get the demand for capacity 'curves' shown in Figure 4.8c. These are $p_{pc} = 7 - (1/200)q_{pc}$, with $0 \leq q_{pc} \leq 1,400$, and $p_{oc} = 5 - (1/200)q_{oc}$, with $0 \leq q_{oc} \leq 1,000$. We can now obtain the demand curve D shown in Figure 4.8c by summing these two equations (which gives us the portion to the left of X), with the portion to the right if X coinciding with p_{pc}. All this is made explicit in equation(s) 4.20 below, where the valid ranges are also given. In addition, since we are adding vertically, set $q_{pc} = q_{oc} = q$. We thus get the system:

$$\beta = 12 - (2/200)q \qquad\qquad 0 \leq q < 1,000$$

$$\beta = 7 - (1/200)q \qquad\qquad 1,000 \leq q \leq 1,400 \qquad\qquad \{4.20\}$$

$$\beta = 0 \text{ otherwise}$$

With β = 4.5 we can solve the first equation in {4.20} to get q = 750. Suppose that we had tried the second equation. Then we would have obtained q = 450, but this 'solution' does not function with the constraint $1,000 \leq q \leq 1,400$. As for the values of p_{pc} and p_{oc}, we have $p_{pc} = 7 - (1/200)750 = 3.25$; and $p_{oc} = 5 - (1/200)750 = 1.25$.

What we have found out here is that the (marginal) cost of capacity, 4.5 (= 3.25+1.25), is allocated between peak and off-peak uses. The total cost for a unit of output is thus 7.25 to peak users (which we get by adding the variable cost), and 5.25 to off-peak users (= 1.25+4). Note: 7.25 and 5.25 are *output* prices.

Now let us see what happens when β = 1. As the reader should verify (algebraically or graphically), this does not work with the first equation of (4.20); but with the second the correct calculation yields q = 1,200. This 'fits' the constraints $1,000 \leq q \leq 1,400$. The entire capacity cost (=1) devolves on peak-load consumers, and if we make the relevant calculation we find that $p_{oc} = -1$. Since negative prices are strictly off-limits, we get $p_{oc} = 0$.

Although off-peak consumers do not have to pay capacity charges, they mus. pay operating costs of $v = 4$. The amount they consume is obtained by simply putting $p_{oc} = 0$ in the appropriate equation: $0 = 5 - (1/200) q$, from which we get $q = 1000$. (This can also be obtained from another equation. Which one?) The total cost of these 1000 units to off-peak users is thus 4 x $1000 = 4000$.

4.4.1 Exercises

1. When q=1 in Figure 4.7b, what are the values of K and L?
2. Complete the right-hand diagram in Figure 4.7c.
3. At A and A' in Figure 4.7d, is there any demand for 'new' (i.e. additional) capacity?
4. What is q" in Figure 4.7b?
5. What *exactly* do I mean when I say "redundant" capital or labor? Discuss using a diagram!
6. In Figure 4.7b, what is q when K=13 and L=30? What is the marginal product of labor (= MP_L) at that point?
7. What is the significance of the line OZ in Figures 4.7a and 4.7b?
8. Assume that $w_K = 1$ and $w_L = 1$. Use these values, together with K* = 10, a = 1, and b=2, to substitute numbers for *all* symbols in Figures 4.7c and 4.7d; and if these values are not sufficient, add some yourself, making sure that they 'fit' into the picture. Among other things, try to formulate linear equations for the two demand functions D' and D".
9. In the numerical peak-load problem, what are the *total* costs for peak load consumers? Determine in two ways! In the discussion of peak load pricing, readers are occasionally asked to verify certain things, and on one occasion to obtain the slope of a linear equation, and some intercepts. Please do all this!

4.5 Conclusions

This has been a long chapter, however that was unavoidable. Natural gas is a subject that is not going to go away as long as the world needs a large source of high quality energy. Serious students of this subject might require more than this chapter, and here I can suggest my earlier book (1987) and the book of Angelier (1994).

Not much has been said here about the environmental aspects of natural gas. I can remember a conference in Hong Kong in which I and many others praised natural gas to the high heavens; but now, in Sweden, when a replacement for nuclear energy is discussed, natural gas is sometimes put in the same category as coal. Apparently there is a hearty thumbs down by some observers on increased levels of methane in the atmosphere.

There is a brief discussion of the environment in the next chapter, and scattered references at other points in the book, but I think that I can say now that where this crucial topic is concerned, the work of economists is just beginning. Very likely it will require more dedication and sheer knowledge that any other subject in economics.

4.5.1 Exercise

1. Explain the following! (i) "Gazprom (Russia) has made it clear it won't give up its market share." (ii) A new gas pipeline (the Interconnector) will allow suppliers to "offer incremental supplies on a spot basis for gas that isn't committed to long term contracts". (From the *Wall Street Journal*, Jan 22, 1997.)

Chapter #5

Some Aspects of the World Coal Market

The age of coal has apparently come and gone, but many observers feel that it will return, and in some cases have even set a date. According to various well-placed observers in France, between 1996 and 2020 about 1,200 gigawatts (= 1,200 GW = twelve hundred billion watts) of new electricity generating capacity must be constructed and installed around the world, and up to 2010 most of this will be fueled by gas or coal. In fact, with the 21st century in sight, coal generates considerably more of world electricity than gas, oil, nuclear, or hydro. After 2020, however, it is possible that coal's domination will increase. According to the world's leading turbine manufacturer, General Electric, over 1,000 GW of new electrical generating capacity will be needed by 2005, and this is only the beginning. Mr Ron Knapp, chief executive of the World Coal Institute puts it as follows: Developing countries "will use coal in the way we in developed countries used coal 100-150 years ago – use it as a building block of economic development. There is an enormous demand for energy".

Some of this is a revelation to a few of us. Coal has always been considered the least flexible of all fuels; the capital requirements of large, coal burning facilities are considerably greater that those using gas or oil; coal handing problems can be extremely complex at every point in the coal supply chain; and virtually every member of the television audience knows about coal's environmental shortcomings.

But at the same time, coal is more abundant than oil or gas. It appears to be more accessible geographically and politically, since it is produced in ample quantities in 50 different countries, and reserves of minable coal can be found in over 100 countries. Electricity is often – but far from always – less expensive when it is coal based than when generated using oil, gas and nuclear facilities; and genuine progress has been made in designing equipment that can burn coal in an environmentally acceptable way. One of the best examples to be found today is Gardanne (France), where coal from a nearby mine is fed into the largest 'fluid bed' installation in the world. In Holland, coal was a strong candidate to supercede natural gas as an energy source at some point before the middle of the 21st century, and as far as I know may still be; while in Germany coal apparently has a very strong sentimental value. Between 1997 and 2005 subsidies of about 4 billion dollars/year will be received by the coal market industry, even though in 1997 the cost of German coal was well above the market price. In fact, South African coal landed in Duisburg was often cheaper that German coal shipped from Ruhr, less than 50 kilometers away. Sentimentality may be the wrong appellation to use here, however, since some politicians have said openly that Germany must have an indigenous energy supply, regardless of the cost.

One way out of this dilemma for the Germans may be the acquisition of foreign properties. In 1999, Ruhrkohle International Mining offered $1.1 billion for the coal mining assets of Cyprus (US). Deals of this sort could transform Ruhrkohle from a ward of the German taxpayers to a major international player in the fast growing international market.

Much of the proposed increased use of coal in Holland and elsewhere will involve the production of coal gas. This is a departure that has been discussed for two decades. The major development in this area seems to be technology of the pressurized fluidized bed combined-cycle (PFBC) type, and early in 1990 one of these installations was opened in a densely populated district of Stockholm where rigorous environmental standards are mandatory. At the present time gasification is generally considered superior to simple burning in a pressurized fluid bed for fulfilling high environmental norms, although it is more costly for electricity generation than conventional combustion processes. Other interesting technologies include coal-water slurries, and burning coal together with water.

It is also possible that someday coal will return to the petrochemical scene, and perhaps via gasification. In the nineteenth century many materials for the chemical industry were derived from coal tar, since coal is rich in the carbon, hydrogen, and other components needed to produce petrochemicals. But once coal gas (or 'syngas' as it is sometimes called) is available, it can – at least in theory – be turned into synthetic oil or methanol that can be used directly as a chemical feedstock, fuel, etc.

The chapter on gas was fairly long, and so by way of compensation this one will be kept brief. In fact, I will only mention here the need to bring 'volume' into the cost picture in a meaningful way, although this concept will be further developed in a later chapter. As readers have already been informed, it is exceedingly important whether a natural resource deposit is large or small. For instance, oil has been discovered all over Sweden, but in order to pay for the equipment to obtain and transport enough of it to keep the discos and student clubs of Uppsala heated during a mid-winter weekend, a large part of this city would have to be mortaged.

5.1 Introducing Coal Supply and Demand

In terms of energy content, the global supply of coal is at least twice that of oil and gas combined. The best endowed regions for coal are North America, Eastern Europe (and particularly the FSU), China, Australia, and Southern Africa. China is the largest producer, followed by the US and India. South Africa, Australia, and Russia are next. Venezuela, Indonesia, and Colombia are in the process of becoming important coal nations. Coal provides about 30% of world primary energy requirements, and perhaps one-half of the world's electricity is coal based. The largest producers of 'hard coal' in 1996 are shown in the following table.

Table 5.1

Country	Output
China	1292
US	849
India	233
Republic of South Africa	197
Australia	196
Russian Federation	156
Estimated World Production in 2000	5300

Source:BP Review of World energy (Various Issues)

Two major coal producing countries that are falling behind are Germany and the UK: German output dropped by 46% and the UK's by 55% between 1960 and 1990. At the same time the US has established itself as the 'swing producer' or the 'marginal exporter': when the price reaches a certain level, the US can supply a very large amount of coal to world markets. (Its supply becomes 'elastic').

Some observers have claimed that the price at which U.S. coal exports become elastic is approximately \$55/t, however this was before the world-wide overproduction that characterized the end of the 20th century. Black coal was Australia's largest commodity export at the turn of the century, with 70 percent of Australia's coal production being sold on export markets for a return of more than 9 billion Australian dollars, and coal from that country has been said to determine the global floor price, although this could definitely change. (In the case of oil, Saudi Arabia and the U.A.E. have been labeled the swing producers, due to low production costs and huge reserves, but Iraq might eventually fit into this category).

The two basic coal types are steam (or thermal steaming) coal, and metallurgical (or coking) coal that is used in the production of steel. Among the major producers (and exporters), Australia produces the largest percentage of coking coal, while South African coal is mostly for raising steam. The major coal exporters in 1996, together with a forecast of exports for 2020 in Mt per year are Australia (49/190) – where 49 is the earlier quantity, and 190 the forecast; South Africa (46/120), U.S. (38/250), FSU (22/?), Colombia (14/100), Indonesia (5/80), and Venezuela (1/40). In 1996, the average CIF (delivered) price of coal to the ARA (Amsterdam-Rotterdam-Antwerp) area was $43/t, of which handling and transportation costs came to more than 50%. Coal loaded into rail cars in the U.S. and Australia presently costs $10-18/t, but ends up costing $40-45/t in Rotterdam after rail, port, and shipping costs are added.

Something that is often 'assumed away' in academic economics is the cost if transportation, but this is extremely important when thinking about coal. In examining the freight market in 1992, the average ARA price of coal was about $40/t, and freight market highs and lows differed by an average of $5. This kind of phenomenon makes the FOB (Free-on-Board) price at e.g. Hampton Roads (Virginia) 'today' a bad estimator of what the CIF price in Rotterdam will be in a few months. Another factor that should be taken into consideration when dealing with freight rates is the type of vessel. Very Large – or so-called 'cape-size' – ships tend to offer better freight rates than 'panamax' carriers.

There is a wide spectrum of coals, and while *peat* is generally regarded as the lowest class of coal, it is often not even considered coal. Perhaps the reason is that it is brown, porous, has a high moisture content, and often contains visible plant remains. Its average energy content is 4000 Btu/pound. *Lignite* (which is about 25% of global coal production) is often called 'brown coal', and while its moisture content is 41%, it has an average heat content of 7000 Btu/pound. *Bituminous* coals, on the other hand are characterized by low moisture content, and the moisture content of a typical *anthracite* coal is only 3%. (Remember, these moisture and energy contents, etc, are *averages*!)

Where energy values are concerned, we distinguish between *subbituminous* coal, with 9000 Btu/pound, and bituminous coal proper with an average energy value of 12,000 Btu/pound. These two categories of coal account for 70% of production. Anthracite coal (which is 5% of production) is jet black and difficult to ignite, with an average energy content of 13,720 Btu/pound. The World Energy Conference specifies two main categories of coal: hard coal, defined as any coal with a heating value above 10,250

Btu/pound on a moisture and ash free basis; and brown coal, which is any coal with a lower heating value. Still another system divides coal into black coal, which ranges from subbituminous to anthracite coal (although in a previous classification subbituminous was a 'brown' coal), and brown coal.

Most steam and coking coal entering into seaborne international trade is high quality bituminous coal. Long distance trade is important at the present time, and will be even more so later in the 21st century; but it is useful to know that it was only because of a sharp rise in steam coal prices in the 1970s, in association with the rise in oil prices, that the large scale export of steam coal by sea became profitable for the first time since before World War II.

The spot markets for coal are only moderately well developed, although in the 1980s it appeared that 20% of world coal trade involved spot transactions, and this figure may still hold. Of course, when average world prices are about \$43/t, and yet some spot cargos are selling for \$30/t CIF Rotterdam, it removes a great deal of the enthusiasm for selling coal. Long term contracts are highly favored in the world coal markets, and there is also the occasional resort to 'captive mines', where the buyer of coal purchases a mine. Coal contracts are mainly written in terms of US dollars, but both Canadian and Australian dollars are sometimes specified as a means of payment.

Before ending this discussion it should be pointed out that while the Btu is the heat unit employed above, kcal (kilo-calories) have become very popular in various coal circles. Fortunately, the conversion is simple: 1 kcal = 3.97 Btu. About 60% of global coal production is used for the generation of electricity, and this proportion is expected to remain stable well into the 21st century. The steel industry takes 13% of coal, and 70% of steel production is dependent on (coking) coal. Roughly, every tonne of steel produced from iron ore (rather than scrap) requires 630 kilograms of coal in one form or another.

Interestingly enough, coal enthusiasts often point out that coal is only 40% of total generating costs at coal fired plants, as compared to gas being 60% at gas fired installations; but this comparison says something about the greater capital intensity of coal facilities, and nothing about the unit cost of electricity.

Acid rain and global warming have become well known terms throughout the world, and are closely associated with the burning of coal. Sulfur Dioxide (SO_2) and oxides of nitrogen (NO_x) are the villains behind acid rain, but these emissions are being greatly reduced due to the technologies alluded to earlier: fluidized bed combustion (FBC), gasification systems such as the Integrated Coal Gasification Combined Cycle (IGCC), and flue gas desulfurisation (FDG). The most annoying problem seems to be Carbon

Dioxide (CO_2), and this is reflected in the increased amount of attention being given to global warming and its possible consequences.

Before going to some technical matters, I would like to suggest that it is impossible to continue to produce the amount of CO_2 that is being produced today without risking grave environmental/ecological consequences. This, by itself, is important, although more important is the gloomy liklihood that if bad news in the form of rising sea levels, etc does put in an appearance, it will already be too late to do anything about it.

5.2 Supply and Cost

Evaluating the supply potential for steam coal, which is the only kind of coal which interests us in this book, involves continuing the descriptive analysis we began above. First we recognize that we have two main types of mines – underground (or deep) mines, and surface (or open pit or open cast) mines. The cost of coal at the 'mine mouth' is determined by the amount of coal in the deposit, the prices of the various factors of production, and taxes and royalties associated with extracting the coal.

For surface mines one of the most important indices of cost is the depth to the coal seam in relation to the thickness of the coal seam. This is measured by the cubic meters of 'overburden' per tonne of coal in the ground. Another important item in the determination of cost is the amount of capital equipment that is used in relation to the size of the mine. Clearly, as the mine size increases, some important economies of scale can be realized. Similarly, for underground mines, the depth of the mine and thickness of the coal seams are crucial determinants of mining costs. For instance, one of the reasons that Western European coal costs so much is the depth of their mines. These often run from 1500 to 4000 feet, as compared to 300 to 600 feet for coal from the Appalachian region of the United States. (1 foot = 0.3048 meter). Soviet mines can also be very deep.

It should also be evident that since most mining output is sold under long term contracts that can run well over 20 years, sufficient reserves must be available to support a given level of production over the life of the contract. This appears to mean that for a surface mine with a 20-year life, one tonne of output per year requires, on the average, 28 tonnes of reserves. This indicates that, on the average 8 tonnes of reserves are unrecoverable. An underground mine having the same effective mine life requires 42 tonnes of reserves *in situ* to support an annual output of one tonne. This is because of the lower ore recovery factor of deep mines.

One of the things that I emphasize in a later chapter of this book is the construction of supply curves that take into consideration *volume* - i.e. the total supply over a certain period – as well as output per period. Employing data from the U.S. Department of Energy, I once constructed some curves of this nature for coal, but the sad truth is that these curves were much less useful than they looked because energy markets are too complicated to be interpreted with such unassuming contrivances. (See however Zimmerman (1981) for an excellent discussion of this matter.) Supply curves, *per se*, suggest that Australian coal has a considerable advantage over U.S. coal, *export*, but this advantage apparently disappears when we factor in freight rates, especially where European markets are concerned. However, at the same time, *backhauling* opportunities can provide significant revenues for Australian colliers (i.e. coal carriers). It has often been possible for ships transporting coal from Australia to Europe to move iron ore or bauxite from Brazil or Africa to markets in the Far East on their return journey to Australia. According to one study, in 1979 an average savings of 6.3 dollars/tonne was possible for a 150,000 deadwight tonne (dwt) vessel that carried coal from Australia to Northern Europe, and returned with a backhaul. ('Deadweight' is the loaded displacement of a vessel minus the empty displacement, and therefore measures actual cargo capacity). Where a sizable backhaul capacity is available, it can often serve to reduce the price at which coal is offered.

In studying the world coal market, it is impossible to avoid noticing the repeated violation of at least two basic tenets of mainstream economic analysis: the frequent failure to exploit low cost deposits of coal before high cost deposits; and the proclivity of some buyers to purchase expensive coal when low-priced coal was available. If we forget about irrationality, then a few of the most important reasons are: imperfect knowledge about present market and production conditions, laws that prevent the unilateral breaking of contracts (and thus keep coal buyers from switching from high to low price coal if the latter becomes available), intertemporal risk aversion based on beliefs about the future price and availability of coal, various political constraints, etc.

It may also be true that large sellers of coal, and their governments, are aware that they are better off if the world market is characterized by the sale of some high cost coal. It is this sale of high cost coal that makes large 'rents' possible for the sellers of low cost coal. The same might be true for oil. There is at least one well known American academic who believes that the world would be a better place if Saudi Arabia flooded the oil market with inexpensive oil – oil that is sold at a price somewhere in the vicinity of the short-run marginal cost in that country. This might mean eventually selling about 18 million barrels of oil per day, but receiving less in total revenue than is being obtained today from sales that are half that size, or less. Frankly, this sounds like a very inauspicious career move on the part of

Saudi Arabia's oil-minister. (The expression 'rents' used above is analogous to profits).

Although it is becoming difficult to avoid the encomiums being heaped on coal by coal industry spokespersons, and even some opponents of nuclear energy, it is not at all certain that this resource is nearly as flexible as oil and gas. Expanding port capacities in order to accommodate the largest and most economical coal carriers can be enormously expensive and time consuming; and clearly, with coal production on its way to doubling in some of the major producing countries, sizable railway investments will be necessary. If these investments are not made, then coal prices will have to leave the range in which they have vegetated over the past twenty years, and they will *not* be drifting downwards. Something that should be expected fairly early in the 21st century is a large expansion in coal trade, since of the 3.5 billion tonnes of coal produced in the world in the middle of the 1990s, only 10% entered world trade.

In Japan expectations are that coal prices must increase, and sooner rather than later. It has been the practice in that country to import only high quality bituminous coal because of the plentiful quantities of this coal that is usually available, but of late utilities have been informed by various research bodies that lower quality coals have an important part to play in their future. Thus, as in the United States, where the burning of soft coal has been acceptable for a number of years, greater attention will have to be paid to the cleaning of coal as well as raising thermal efficiencies. Of course, in Colombia and Australia, it is often claimed that huge savings in production costs can still be made. In both these countries, free-on-board (FOB) port costs are now thought possible which, in the light of expected prices, could mean attractive profits.

The real test for coal in Japan and other countries will come when 'safe' nuclear equipment is available in industrial sizes (i.e. from about 900 MW). If this equipment does not cost more than that being used at present, it hardly seems likely that governments having the same attitude toward energy independence and economic growth as the Japanese government would be interested in coal. In fact, if large scale, safe nuclear equipment did become available, an increasing amount of coal would be 'dedicated' to the production of motor fuel. Despite rumors that are in circulation in France, and elsewhere, it is far from certain that coal will eventually replace nuclear energy in the generation of electricity.

5.3 Some Theoretical Aspects of Coal and the Environment

One of the purposed of this book is to introduce readers to the rhythm and vocabulary of energy economics. This includes helping to make it possible for them to know which topics are relevant, and providing them with some of the theoretical background that is required to ask pertinent questions about those topics. To my way of thinking, game theory has achieved too prominent a role in economics, but even so there is no denying its many attractions. One of its major attractions is the part played in its development by John von Neumann who – in his prime – was often referred to as 'the best brain in the world'. Given von Neumann's wide range of interests, which included the organization of the Roman army, and the American Civil War, this may well have been true. Accordingly, when the best brain in the world takes an interest in something, even the very limited interest that von Neumann took in game theory, then the rest of us have no choice but to sit up and take notice.

In von Neumann's book, co-authored with Oscar Morgenstern (1944), there was no explicit mention of prisoners dilemma games of the type that will be given some attention in this section, although a few of us believe that the reasoning behind the zero-sum type games that were the major theme in their book was just as important for several of the topics being taken up in this book. What this reasoning means to me when considering environmental (and other) matters is SAFETY FIRST! It means ruling out high-risk activities as long as other choices are available. "You know that the best you can expect is to avoid the worst", Italo Calvino once remarked, and according to William Poundstone (1992), this epigram neatly summarizes the central paradigm of 'The Theory of Games and Economic Behavior', which is the widely known but not always well understood *minimax principle*. This can be put another way: the optimal strategy for von Neuman-Morgenstern type games turns on avoiding losing rather than aiming to win. There is nothing wrong with winning, but such ultra-macho behavior as going after an incontrovertible win when there is a high probability of 'ruin' is not recommended. (As an aside, another 'Central European' economist, Erich Röpke, called game theory 'Viennese coffeehouse gossip', and the Princeton mathematician Paul Halmos once announced that he was supremely unimpressed with the subject. The opinion here is that even if game theory is overrated, it contains a number of useful insights, and the elementary game theory literature deserves to be studied by *everybody*, and not just economists.)

The basic quandary that we have to deal within regard to the burning of coal (and the other fossil fuels) is that the atmosphere is communally owned.

If a country (community) decides to maximize the value of its communal rights, then it might overpollute the atmosphere because some of the costs of doing so will fall on others. The capacity of the atmosphere to absorb further pollution might in this case be diminished too rapidly – which suggests that everyone, or almost everyone, could gain if pollution were decreased. Thus it may be so that the owners of the atmosphere (i.e. all countries) could determine and agree on an optimal pollution level if they entered into negotiations with each other. Optimal in what sense? In the sense that fossil fuel use and economic growth go together at the present time, even though too much pollution via fossil fuel use could diminish the positive welfare effects that are sometimes associated with growth. (With economic growth it is possible to finance increased health care, but pollution has a negative effect on health.) In the conference held in Kyoto (1997), concepts of this nature were frequently expressed. Of course, in order to ensure the implementation of international agreements, some threats may be necessary, along with the power to carry out these threats. This kind of observation is not common in the environmental-ecological (e-e) literature, and may not be especially popular with persons who produce that literature, but once e-e concerns come to the attention of students of game theory, reference to the formulation and carrying-out of threats are not easily avoided. (Incidentally, some of the e-e people have a vocabulary that would fit in well with the formulation of threats. For example, a member of Ozone Action, referring to the proposed merger of two giant oil companies, called it the "Death Star of global warming".).

We can begin our discussion by making sure once again that we know what *equilibrium* means. Mainstream economics is, for the most part, equilibrium economics: we study equilibria, and *not* the details associated with the movement from one equilibrium to another. In the classroom I have often used a definition of equilibrium taken from physics – a state of rest; but this would hardly satisfy many formalists. He or she would insist that in economics equilibrium means *correct expectations*: a situation where the outcome of a transactor's attempts to perform a certain action – such as sell a given amount of a particular commodity, or service, or financial asset at a particular (e.g. market) price – conforms with her expectations, although in this context 'intentions' might be a useful word.

Now let us look at a prisoner's dilemma type game, beginning with the original anecdote. Ms Bibi Sally and Professor Bill Lather are picked up on suspicion of having robbed a parking meter, taken to the local police station, and interrogated separately. If one of them confesses (C) and the other denies (D), then the one confessing receives a light sentence, while the one denying receives a very heavy sentence. If both confess the sentence is moderate; if neither confess, they are given a month in jail for loitering. The *normal* (or matrix) form of the game is given in Figure 5.1a, with Bibi's sentence first, and the lengths of sentences in months. For example, if Bibi denies and Bill confesses, she is given 12 months and Bill ½ months.

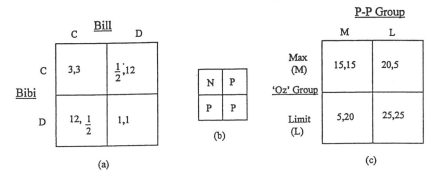

Figure 5.1

The dilemma in all this is that if they trust each other, and via the medium of this trust cooperate, they play (D, D), and receive light sentences. Isolated, however, cooperation seems difficult to justify, and this leads each of them to confess. The 'solution' turns out to be (C, C). This is called a Nash Equilibrium – actually a Nash-Cournot equilibrium, since in reality it is John Nash's formalization of the equilibrium presented by Antoine Cournot more than a hundred years earlier. It is an equilibrium since if both parties play it, a unilateral defection would be ineffective: C is the best play for Bibi if Bill plays C (and also if he doesn't), and vice versa. A better outcome for both is (D, D), but if they played this they would be tempted to defect, and the assumption is that the temptation would be overwhelming.

It does not have to be said, I hope, that this last assumption depends on the values in the 'boxes'. If the sentences shown were measured in days instead of months, then there is a good chance that neither would consider confessing, since there is a certain status to being regarded as a 'stand-up guy' by the street crowd, and to attain this distinction you cannot make a practice of confessing.

Next we take these concepts to the pollution arena. The assumption in 5.1c is that we have two groups of countries: rich countries, labeled the 'Oz Group', which holds most of its meetings in Sydney (Australia), and the Pago-Pago (or P-P) Group, which holds most of its meetings in Monte Carlo, since Pago-Pago does not have a suitable conference venue. If the Oz group of countries *and* the P-P group decide to limit domestic pollution (L, L), both realize the utility (or satisfaction) level 25. This level takes a number of things into consideration, to include the global benefit of pollution abatement, less harassment from Greenpeace, intense feelings of moral superiority, etc.

But if both continue on their present course of maximizing personal satisfaction, then they pay a penalty in terms of higher pollution levels for both, as well as having to anticipate even higher levels in the future – which is something that could result in considerable mental and physical distress. The remaining 'plays' involve one party undertaking a serious pollution abatement program, while the other group continues to go all out in the way of generating pollution.

The Nash Equilibrium is (M, M), corresponding to the (C, C) solution in Figure 5.1a. Once again the assumption is that the cooperative solution, (L, L), is unstable, in that cooperative agreements, if reached, would not hold. If we were talking about property that was closer at hand than the upper atmosphere, for example the oceans, then a way out of this predicament might be to establish boundaries, and therefore private-property rights to anything within these boundaries. There are well known propositions in economic theory which claim that the establishment of these boundaries would lead to a global optimum.

Once we are prepared to deviate from prescriptive economics, however, we might be able to do something with our atmospheric pollution problem. As will be noted in the next section, countries *do* enter extremely significant international agreements, although the television audience is not made adequately aware of this fact, and what it means for their future, as well as the future of their descendents. This being the case, if the *right* countries enter into these agreements, they might be able to put enough pressure on countries that do not want to join in to bring them to their senses. Good examples of this are the international agreements on nuclear disarmament – or partial nuclear disarmament; and, on another level, the Packwood-Magnusson amendment to the U.S. Fishery Conservation and Management Act (1976), which requires the U.S. government to retaliate whenever foreign nationals overstep their fishing rights. An offending nation automatically loses half its allocation of fish products taken from U.S. waters, and is given a year to change its behavior. If it refuses, its right to fish in U.S. waters is automatically revoked. Presumably, this kind of language means that gunboats and helicopter gunships are somewhere in the background.

Before looking at another portrayal of this issue, the reader should examine Figure 5.1b, where the P's designate non-equilibrium situations. However these are interesting non-equilibria – interesting and known to almost every survivor of the introductory course in economics. Situations where one player can better his or her prospects only at the expense of another. The cooperative situation, (D, D) = (1,1), is a Pareto optimum, since if e.g. Bibi moves to C, she 'walks' in two weeks, while Bill has 12 long months of hard time on the chain gang to look forward to.

Another interesting ways of putting it is that one play (or act or policy) is better than another only if *every* individual agrees that it is preferable. That leads us to a concept used often in game theory, where we say that if a feasible act cannot be upset by the grand coalition of all individuals, then it is Pareto efficient. (C, C) is *not* Pareto efficient, because if Bibi and Bill could form a 'grand coalition', they would upset this arrangement by moving to (D, D). Similarly, they would not form a coalition to upset (D, D). Unfortunately, due to the separate interrogation clause in the prisoner's dilemma game, they are not given the chance to form coalitions.

By way of continuing, a trivial example from game theory will be introduced in order to show readers how backward induction functions. This is an extremely important concept, and students should find out something about it as early as possible in their basic training. Suppose that one group of countries, the Oz group, considers threatening another group, the P-P Group, because the latter did not sign an agreement to comply with some sort of pollution abatement program. The threats might involve comprehensive economic sanctions or, worse, not inviting political and economic representatives of the P-P Group of countries to lunches and cocktail parties. Without going further then the threat stage, Figure 5.2 might represent this situation. T is threat, NT is no-threat, C is compliance, while NC is non-compliance. This is the *extensive* (or game-tree) representation, which emphasizes the dynamic nature of games. First Oz moves, and then the P-P Group. The outcomes of this game are shown in the next to the bottom row, and each outcome has been given a numerical value. These are shown in the last row, with the 'utility' of Oz's outcome shown on top. Some attention can also be paid here to the *nodes* A and B. These are independent in the sense that when the P-P Group makes their move, they know whether they are at A or B: whether a threat has been made or not.

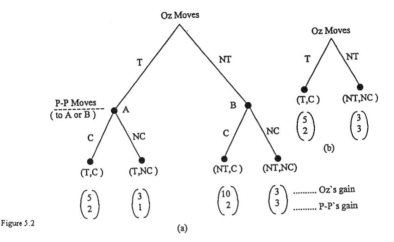

Figure 5.2

(a)

(b)

We are now in position to get a useful insight into what backward induction is all about. If Oz plays NT, then P-P plays NC. Why? Because if P-P is at node B, playing NC provides a gain of 3, as compared to the gain of 2 if C had been played. It's that simple! By the same token, if Oz had played T, then P-P plays C. The reason is the same as before: 2 is preferred to 1.

Now that we have settled P-P's choices, depending on whether this group finds itself at node A or node B, we can turn to Oz's behavior. If Oz plays NT, they will have to face the outcome (NT, NT), which to them has a value (gain) of 3; while if they play T, they are confronted with the outcome (T, C), which is worth 5 to them. Obviously, T is the correct choice, and the solution of the game is (T, C), with a value of 5 to Oz, and a value of 2 to the P-P group.

There is hardly any point in making heavy weather of this example, but perhaps the reader should notice that non-compliance (NC) has a higher value for P-P in the absence of a threat than with a threat. A reason for this might be the understanding by the power-structure in P-P of the bad news that could emerge on the lunch/cocktail-party circuit if Oz's threat was not taken seriously.

The final concept that will be presented here is the *core*, whose original depiction should be attributed to Edgeworth (1881). The core is particularly useful when trying to systemize bargaining theory. According to John Nash (1950), it is necessary to emphasize that if an agreement cannot be reached, then the players remain in possession of what they started out with. He should have added, at least temporarily. Thus the failure to reach an agreement is a kind of threat, particularly since an agreement – if reached voluntarily – represents a clear gain, though perhaps not the gain you desired, and thought was possible. In addition, if you do not lose from not entering into an agreement, then your original position is a kind of security level.

The kind of agreement that we are thinking about here should be such that it cannot be bettered by some other agreement that is realizable, given the mind-set and behavior of the person or persons on the other side of the table; and it cannot be bettered from the point of view of any negotiator by leaving the table. If an agreement meets these criteria, then it is in the core of a game. Clearly, there are many different agreements possible here, which lends an annoying indeterminacy to all this. We end up with a basket full of solutions, or possible solutions; and although many real life bargaining encounters conclude with papers being signed and champagne glasses raised in front of smiling faces, the mathematical interpretation of bargaining games in the seminar room inevitably entails some kind of arbitrariness in choosing solution.

We can continue with an example in which two groups of countries, for example Oz and P-P, can abate a larger amount of pollution by working together than is possible if they work separately. The gains from working separately are shown as G_o and G_p, while the gains from working together are found on the feasibility frontier in Figure 5.3a. As implied by our previous discussion, a portion of the feasibility frontier is uninteresting. For instance, to the left of A and to the right of B is the core. The stretch between A and B is the core. The bargaining game is about where the bargainers are going to end up on this stretch.

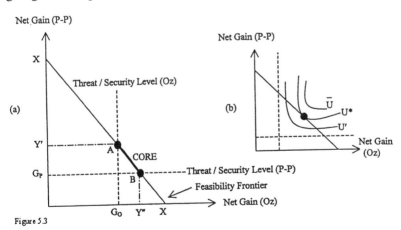

Figure 5.3

As pointed out above, theory tells us how to locate the entire core, but not the thing we are most interested in: where in the core the players will end up. To find this out some assumptions of an arbitrary nature are necessary. One possibility is to divide up the total amount they would possess if they worked together, X, in the same proportion as the original distribution:

$[G_o/(G_o+G_p)]X$ for Oz, and $[G_p/(G_o+G_p)]X$ for P-P. (Exercise: Show how much these shares add up to!).

An argument against this arrangement might be that Oz has a great deal to begin with, and if P-P forwarded this argument vehemently enough, Oz might be inclined to listen. As it happens, John Nash has an interesting – though hardly revolutionary – approach to this matter. He derives a rather special system of utility curves (as in Figure 5.3b) and uses them the same way you used similar curves in your first course of economics. The solution is at the point where these curves are tangent to the efficiency locus. Exactly how useful this technique would be in real-life negotiating situations is something the reader must determine for himself and herself, and preferably before negotiations began.

That brings us to the end of this short theoretical discussion. Some readers may wonder why it is here, but I do not see how it can be avoided. During the Cold War (and the war in Vietnam), the Rand Corporation (of Santa Monica, California), lavishly funded by the U.S. government, used concepts of the above nature to study the optimal moves of potential opponents, and how these moves should be countered. Some of their solutions were labled absurd – or worse – by both the scientific and military communities, but even so the same kind of exercises, better thought out, might have a future if they deal with energy, and environmental and/or ecological topics. Perhaps you might be able to contribute to projects of that nature? If so, the more you know about the methodology introduced in this section, the more you will have to offer.

5.3.1 Exercises

1. The end points of the feasibility curve in Figure 5.3a both have the value X. Suppose they were X and Y, with $X \neq Y$. Suggest a method of dividing up any gain obtainable due to Oz and P-P working together.
2. Discuss nuclear armament and disarmament – India vs. Pakistan – using game theory.
3. Change the numbers in Figure 5.2 so that the solution to the game turns out to include non-compliance on the part of P-P. Discuss in detail!
4. Are core solutions Pareto optimal solutions? What about a solution at G_o, G_p?
5. In the last paragraph of this section I say that eventually some studies might have to be carried out of the type originating with the Rand Corporation, only this time dealing with environmental and/or ecological matters. What am I implying?

5.4   Coal and CO$_2$

A major concern at the present time is with the release of CO_2 that accompanies the burning of coal, and the global warming/Greenhouse-effect that it could eventually bring about. As touched on earlier in this chapter, the technology exists for obtaining at least a partial solution to the problem of atmospheric pollution, although whether this technology will be utilized to the optimal extent remains to be seen.

The level of CO_2 in the air (i.e. the 'lower atmosphere') increased by at least 15% over the period 1865-1995, and if this rate of increase continues, then radical adjustments in the climate are theoretically possible. I would be more than pleased to summarize some of the changes that have been publicized, however a number of academic economists, and others, have become more aggressive than ever in injecting a spurious precision into the environmental debate. As emphasized in a pronouncement of the United States Environmental Protection Agency (EPA) some years ago "...the mathematical and statistical models used to forecast ozone depletion (due to such things as the use of chloroflurorcarbons) are so inaccurate that their use is probably counterproductive." (Quoted in Boxall (1991).)

The complication with global warming is the well known fact that even small changes in average temperature can have dramatic consequences. Around the year 1300, with the temperature in the Northern Hemisphere at about the same average level as today, a fall in the average temperature of about one degree centigrade was enough to make ice skating on the Thames possible during a few weeks of the winter. Similarly, it has been estimated that a one degree (average) rise in the temperature might cause an unwelcome increase in the amount of bacteria in the surroundings. In the last 100 years the average global surface temperature has risen by about 1 degree Fahrenheit, to nearly 60 degrees Fahrenheit (= 15.6 degrees centigrade), with the 1990s being the warmest decade of the 20th century. (This, by itself, does not have a great deal of significance). Estimates now are that the average global surface temperature will rise by another 3.5 degrees in the 21st century, unless heat trapping industrial gases like CO_2 can be reduced.

Without becoming unduly alarmist, it needs to be stressed that our civilization is much more finely-tuned to the present climate than the television audience has been led to believe. For instance, an increase in the average annual temperature by 3 degrees for the entire globe, would mean up to 10 degrees in the polar regions during certain periods of the year, while at the same time there may be no change at all in the vicinity of the equator. But it is the pole-to-equator temperature difference that determines wind patterns around the globe, and these in turn determine how much rain falls, as well as where it falls. Changing pole-to-equator temperatures by 10 degrees or more could lead to a geographical reallocation of agriculturally productive land that, together with irregular changes in productivity, would result in a lower global agricultural output.

Inadequate agricultural output at some point in the future is already a threat, even without having to contend with global warming, and given the rate of growth of world population, it would be a universal tragedy if there was a reduction in the output of agricultural surplus areas such as the United States.

Recent research suggests that warming may lead to more frequent appearances of El Nino, the huge pool of warm water in the eastern tropical Pacific Ocean that occasionally ignites an extended period of meteorological destruction. (The cold water counterpart is La Nina.)

Scenarios of the above type are the reason you found conflict-oriented models in this chapter. This is not to propose that the Texas Rangers or Nato – augmented by the Former Soviet Union – should become the gendarmes of the upper atmosphere, but to suggest that in addition to making a joint commitment to environmental sanity in order to influence allies and wellwishers, the superpowers, near superpowers, and potential superpowers should attempt to judge and eventually takes measures to counter what may amount to a lethal threat to our environmental/ecological heritage. For instance, the increasing CO_2 levels noted above could unleash substantial amplification (or multiplier) effects. Rising soil temperatures might escalate the production of CO_2 because the decreasing alkalinity of saline seawater will decrease the solubility of carbon dioxide at or near the surface of the water.

Some economists have approached the carbon dioxide impasse as a complex problem in welfare economics that, at least in theory, could be solved if persons of good will were prepared to put their trust in some elementary results from the calculus of variations, or its modern version, optimal control theory. Unfortunately, however, as Jean-Claude Charrault of the Directorate-General for Energy of the Commission of the European Communities once put it "…the decisions we take now, and those we have already taken, will affect our children, grandchildren, and great grandchildren for many generations. We've got to get it right!"

Getting it right means a direct attack on such hazards as acid rain and the greenhouse effect that takes the form of prohibitions and regulations, and not cute algebraic demonstrations with only a superficial application to real-world perils. It means the setting of environmental standards, the honoring of these standards, and in addition governments putting pressure of one sort or another on governments that violate these standards.

Suddenly I conclude that the previous discussion can be summed up as follows: 'Standards. No compromise'!

Can this be done? In May, 1990, when an international panel of scientists warned that if the discharge into the atmosphere of large amounts of such things as CO_2, methane, etc were not immediately reduced by a substantial amount, global temperatures could rise in such a way as to have unforseeable – but dangerous – consequences for the entire human race, the UK Prime Minister, Mrs Thatcher, dissociated her government from the skepticism of the US administration as to the need for immediate action, and stated that if all governments did their part, the UK would attempt to stabilize its carbon dioxide emissions at the 1990 level by the year 2005. Subsequent intergovernmental meetings, such as those related to the Montreal Protocal (which had as a goal the reduction in ozone depletion), made it clear that a new era of international cooperation was not only possible, but likely. Environmental groups in the US, as well as the UK House of Commons Energy Committee, have recommended setting an example to the world by tackling the domestic emission problems of their countries *in advance* of international action. Estimates of the magnitude of ozone depletion (in percent) provide a cogent example of the remarkable results that concerted action might bring about between the years 2000 and 2075. (Note: *might* and not will, because an accurate calculation is patently impossible)!

Table 5.2

Case	2000	2025	2050	2075
No Controls	1.0	4.6	15.7	50.0
Montreal Protocol	0.8	1.5	1.9	1.9

Source: EPA(US), 1988,

The issue with ozone depletion concerns chlorofluorocarbons (CFCs) and halons, but we are faced with the same type of perplexity when considering the burning of coal. For example, had coal been introduced into the world economy at the rate at which the International Energy Agency (IEA) wanted it introduced in 1974-75, then carbon emissions from the generation of electricity would be more than 50 percent higher than they are at present.

One of the great shortcomings of academic economics is the constant search by its practitioners for new subject matter – not for new solutions, but for new motifs. Of late economists have started playing laboratory games involving such things as small sums of money and 'affection', somehow implying that these charades will give an insight into real-life wheeling and dealing on Wall Street, or negotiations about political and economic matters at the Palais des Nations (Geneva). Without going too deeply into this matter, let me suggest that so-called experimental economics has only a limited application. The best way to launch an effective program against environmental deterioration is to make it a part of a general scenario for the reduction of international tensions and inequities. (It has turned out that making experimental economics respectable has led to the emergence of something called behavioral economics. What this turns on is the belief that any behavior that can be described verbally, can be put into mathematical form. Thus, no assumptions are necessary about such things as rationality. Aside from the fact that there is already too much mathematics in economics at the graduate level, it would be impossible for me to imagine any reputable mathematician having anything to do with this kind of thing.)

Something else that might be fruitful is to get politicians and civil servants dealing with environmental matters to learn and apply a fundamental proposition of game theory: whenever cooperation is possible, then almost without exception it is in the interests of all players to see that it takes place.

One last observation. A 500 MW power station in the US burning coal produces annually, on the average, 5000 tons of suphur oxides, 10,000 tons of nitrogen oxides, 500 tons of particulate matter, 225 pounds of arsenic, 4.1 pounds of cadmium, 114 pounds of lead, plus various trace elements. The average coal input is 1.5 Mtons, which is transformed to ashes, as well as, roughly, a million tons of CO_2. The environmental damage done by a power plant of this size is difficult to estimate, but it is not a good idea to pretend that it is insignificant. Needless to say, power plants of the same size in most other parts of the world are unlikely to be more attractive from an environmental point of view.

5.5   Conclusions

Danger invites rescue
-Benjamin Cardoso

In a few decades coal may become as important an energy topic as oil is today. This, at least, is the opinion of many observers, and to a certain extent they could be right.

But to my way of thinking, coal is not the ultimate backstop, which it has occasionally been called. Far from it. Like natural gas, it is a transitional energy medium. Eventually, renewable energy resources are going to be required, and required in large amounts. Something else that will probably be required is huge amounts is electricity from fission or fusion reactors, and this is going to be true regardless of the progress made with wind, wave, and bio-energy.

Such being the case, the duty facing readers of this book, and their friends and neighbors, is to make sure that this nuclear equipment is the safest possible; and that no tolerance is shown unsafe facilities or operating procedures. The thing to remember here is that there are various substandard nuclear installations in full operation in this old world of ours, and in the course of a vodka bash in the control room of one of them, somebody might press a button or pull a switch which leads to an accident that sours the electorate on nuclear energy forever. Let me also use this opportunity to impress on various enemies of nuclear power that a low energy society will be one that, at least to begin with, will be characterized by constant social and political turmoil, as its economy transmutes toward a new equilibrium. Eventually, of course, after the smoke has cleared, it might become more tranquil than ever. But then again, it might not,

Christian Gerondeau once concluded that energy crises were *passé*, and that in the near future – and forever after – there would be a surplus of energy. What there will be is a surplus of optimistic books and learned papers to show that, with a little luck, a post-industrial paradise can be created out of such activities as 'surfing on the internet'. The truth is somewhat more dismal. In a world where population could reach 8 or10 billion human beings before the middle of the 21st century, the real energy crisis may be about to begin.

5.6 APPENDIX: SOME TRANSPORTATION ECONOMICS

Some of the persons reading this book will write economics books and articles of their own some day. It is both easier and more difficult than you think. If you diligently train your concentration and patience, the actual writing might be easy; but you will not be able to avoid having to select what to include and what to omit, and very often this is not easy. In this chapter, for example, I decided to include a few topics that are extremely important, but which you would not expect to find in an elementary textbook.

By the same token, in this appendix, I present some algebra that you would expect to find in a book on coal, but which is usually absent. It has been said that there is a disinterest in sub-bituminous coal on international markets, which may be true, but this contention is misleading to a certain extent – usually because only a single transport cost is considered. In reality this cost can differ because of vessel size and characteristics, route, possibilities for backhauling, etc. Taking P as the coal price, and θ as the heat content (in Btu), the equilibrium condition for coal from sources x and y – considered with respect to a third country – is:

$$P_x/\theta_x = P_y/\theta_y \tag{1A}$$

134

If $\theta_x < \theta_y$, then we must have $P_x < P_y$, which simply says that the coal with the lowest heat content must sell at a lower price. Now introduce total transport costs (from mine-mouth to market) of t_x and t_y. This gives us:

$$(P_x+t_x)/\theta_x = (P_y+ t_y)/\theta_y \qquad (2A)$$

Solving for P_x we get:

$$P_x = \frac{\theta_x}{\theta_y} P_y + \theta_x \left[\frac{t_y}{\theta_y} - \frac{t_x}{\theta_x} \right] \qquad (3A)$$

Note that $dP_x/dt_y > 0$. (Or, if you want, think of this as $\Delta P_x/\Delta t_y > 0$.) Thus, if the low Btu coal has a much shorter journey to market, its price, *ceteris paribus* need not be as low as is sometimes thought. Furthermore, if we note that most of the coal in seaborne trade has a similar Btu rating, then we immediately see the importance of transport costs. If $\theta_x \to \theta_y$ (in the limit), then we get

$$P_x = P_y + (t_y - t_x). \qquad (4A)$$

The simple message presented here is that we can have P_x greater than, equal to, or less than P_y; and since the interest shown a commodity (such as coal) in most markets has to do with price, then we see from the last expression, or from 3A, that even low Btu coal might be able to find a place in the international market (assuming, of course, that their environmental shortcomings are the same).

Similarly, although it is not always intuitively evident, falling freight rates can increase the competitiveness of coal whose mine-mouth and/or FOB price (F) plus the transport cost (T) is relatively high. To show this, begin with a situation where the delivered price of coal is higher to a third market if purchased from country X than from country Y, with the delivered price defined as F + T. The statement of the problem then gives:

$$F_X + T_X > F_Y + T_Y \qquad (5A)$$

The assumption is also made that $T_Y > T_X$, although $F_X > F_Y$. (It is this assumption that keeps the exercise from being too trivial.) Next, suppose that there is a uniform reduction in transport costs for both countries of λ percent. Costs are now $F_X + T_X - \lambda T_X = F_X + T_X(1 - \lambda)$, while the same operation for Y will give $F_Y + T_Y(1 - \lambda)$ as the new cost. The question that will now be asked is what value of λ is necessary to reduce the delivered price of X's coal to a level that is lower than the price of Y's coal. Symbolically this means that $F_Y + T_Y(1- \lambda) > F_X + T_X(1- \lambda)$. Some elementary manipulation will allow this to be rewritten as:

$$\lambda > 1 - \frac{F_X - F_Y}{T_Y - T_X} \tag{6A}$$

To see how this works, we can use some approximate values for Australia (X) and the US (Y) during a short period at the beginning of the 1980s, with Western Europe as the third market. With $F_X = 52$ and $T_X = 27$ for Australia, and $F_Y = 58$ and $T_Y = 15$ for the US, the inequality (5A) gives $\lambda > 0.50$. In other words, if ocean shipping costs fell by more than 50% for Australian coal, it would be cheaper in European markets than US coal. This is exactly what happened.

Chapter #6

Energy Derivatives: Futures, Options, and Swaps

Although this chapter is called 'Energy Derivatives', all the examples involve oil. But by the same token, the title could have been commodity derivatives, because the elementary theory presented below is valid for such things as copper and aluminum, as well as e.g. oil and gas. It also applies to a considerable extent to the evolving markets for electricity derivatives; and it makes an excellent introduction to financial derivatives.

I have endeavored to provide a fairly complete introduction to energy futures and swaps, but space considerations made it necessary to leave out a few important topics in options theory. At the same time, however, most readers should leave this chapter with an adequate knowledge of what options are and how they function. In fact, the time and effort spent absorbing the information in this chapter could turn out to be one of the best investments you can make.

An informal introduction to futures can be found in Section 3.4 of Chapter 3. It may useful to examine that discussion before continuing.

6.1 Futures Markets: Terminology and Nomenclature

If the reader had no problem in following the informal presentation of futures markets in Chapter 3, then he or she already has a useful insight into the logic of a market that seems to be misunderstood by all except a few specialists. Let me also note that in the ensuing discussion, the algebraic and graphical techniques employed do not go beyond those used in introductory economics courses, except for the summation (i.e. Sigma) signs; but an elementary explanation of this notation can be found close to the beginning of Chapter 2. If, however, the reader elects to skip the brief algebraic-graphical analysis in this section, then he or she should definitely not ignore the discussion of basis risk that comes later. Basis risk had a very important part to play in giving Metallgesellschaft (MS) its 1.5 billion dollar headache, although simple bad luck played a salient role.

At the center of futures markets we have the futures contract, which can be considered a highly specialized forward contract. Against a background of speculators 'betting' on the direction and size of commodity price movements by buying and selling futures (and options) contracts, it is possible for producers, consumers, inventory holders and other traders in physical products to reduce (and in some cases eliminate) undesired price risk by also buying and selling these contracts. More specifically, traders in a physical product can employ futures markets to reduce risks if other traders and/or speculators are willing to assume this risk. A principal source of social gain from futures and options trading derives from the voluntary redistribution of risk *from* risk averse transactors in physical products *to* speculators. The directors of futures markets are prone to insist that these institutions exist only to provide price insurance for producers or consumers of 'actuals' (or 'the underlying'), but clearly speculators are extremely important. (Theoretically, they are not indispensible, but a futures market without a heavy input of speculation probably does not have a bright future.)

If a speculator believes that the price of a commodity is going to rise, he buys futures contracts for that commodity. (He might, however, buy futures for a similar commodity, which is one way that *basis risk* comes into the picture.) Traditional futures contracts are also forward contracts, since delivery conditions are specified on them relating to a specific amount of a commodity, delivered during a certain period to one or more specified locations; but it is possible to avoid taking delivery if, at any time before the contract matures, an offsetting (i.e. reversing) sale is made of a contract for the same amount of the commodity, referred to the same delivery period. (Usually beginning at the end of a month.) As a rule, less than 5% of futures transactions involve delivery. In the 1980s contracts with cash settlement (instead of delivery) were introduced, where traders' gains and losses were calculated from a known price, or price index. Losers then pay winners.

Offsetting – or closing an open position before the maturity date of the contract – is always possible in a viable futures market. Here it should be noted that a viable market is one with a high degree of liquidity, resulting from the presence of a large number of active transactors, where transactions can be carried out very soon after buy or sell orders are given, and at a price close to that of the last recorded transaction. If a contract is bought, and sold later at a higher price, then the (long) speculator has made a profit. Similarly, if a speculator thinks the price is going to fall, then he or she opens the position by selling a contract (i.e. going short), hoping to make a profit by closing the position at a lower price.

Hedgers also buy and sell futures contracts, depending upon whether they want to guard against price rises or price falls. Consider, for example, someone who has contracted for a given quantity of crude oil, but does not know the price at which this oil will be delivered because the seller follows the familiar practice of charging the *spot* price of oil at or around the delivery date. The buyer of crude oil thus faces considerable price risk in that the price of oil might rise sharply; but since the price of futures contracts and physical oil should be fairly close at the time the buyer contracts for future deliveries, risk averse buyers can 'lock in' a price for the future by initially buying futures contracts (going long), while making an offsetting sale around the time the oil is delivered. If the spot price of oil rises, the buyer takes a 'loss' on the physical transaction; but (as will be shown later) since the price of futures contracts should also rise, a compensating gain will be made on the (offsetting) sale of futures. The price that is locked in is that relating to the futures contract at the time the transaction is initiated.

Similarly, if a firm is selling oil, and is afraid of a price fall, then hedging takes place by *selling* futures contracts. If the price of the physical oil does fall, then the price of futures contracts should also fall, and a compensating gain will be made via an offsetting transaction that involves buying futures. (Gain = selling price *minus* buying price.)

Now let us move to a simple model that illustrates some aspects of the above discussion. It features a market in which, to begin with, there are only speculators buying and selling futures contracts, of a single maturity, in the presence of an 'auctioneer' who varies the price in such a way as to (eventually) clear the market. This is analogous to the 'open outcry' system that was once employed by most organized exchanges, and is probably the basic form of an auction market (although it has been modified somewhat by computerized trading). Although individual speculators have different opinions about the future of the commodity (e.g. oil) price, their behavior is such that when the price of futures contracts is what they regard as high, they will sell these contracts, hoping that a profitable offsetting purchase can be made later at a lower price; and when the price of futures contracts is relatively low they buy, hoping to make a profit on a price rise. This point applies to the curve D_s-S_s and the curve D_{sj}-S_{sj} in Figure 6.1. Readers should do everything possible to understand the contents of this figure.

Warning! If you have some problems with this figure, or with the algebra in the discussion below, then go directly to the paragraph in this section that begins with 'Some elaboration on the expressions...' The rest of this chapter can be understood without an understanding of these particular algebraic materials, and what follows after this algebra is too important to miss.

140

Consider the (excess) demand $E = D - S$ for speculator 'j', in Figure 6.1b. At $\overline{P}_j$ there are no sales or purchases. To the left of $\overline{P}_j$ this speculator is selling contracts, and thus $E<0$; while to the right of $\overline{P}_j$ they are being purchased (but not sold), and so with $E = D - S$, we have $E>0$. Now consider the *aggregate* (excess) demand curve D_s-S_s shown in Figure 6.1a. To the left of P_f *aggregate* excess supply is positive, although some speculators may be buying rather than selling contracts. (To grasp this point, the reader should construct an aggregate curve for two speculators.) Given the D_s-S_s curve in a market where we *only* have speculators, the equilibrium price is $\overline{P}_f$, with the supply by some of the speculators being equal to the demand by others.

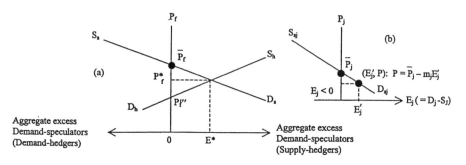

Figure 6.1

Also shown in Figure 6.1a is an aggregate supply-demand curve for futures contracts by hedgers (S_h-D_h). The behavior of hedgers will be explained later in this section, but the supply and demand curves for *individual* hedgers are shown in Figure 6.2. Once again it might be useful for readers to construct an aggregate supply-demand curve from the individual supply and demand curves of hedgers, making sure to observe the following: in order to get an aggregate supply-demand curve that is continuous at the vertical axis, a great many individual curves would probably be necessary. However even if the aggregate supply-demand curve for hedgers is not continuous at the vertical axis, nothing would be changed in the following algebraic demonstration.

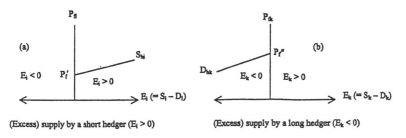

(a)

P_{fi}

S_{hi}

$E_i < 0$ P_f'

$E_i > 0$

$E_i (= S_i - D_i)$

(Excess) supply by a short hedger ($E_i > 0$)

(b)

P_{fk}

P_f''

D_{hk}

$E_k < 0$ $E_k > 0$

$E_k (= S_k - D_k)$

(Excess) supply by a long hedger ($E_k < 0$)

Figure 6.2

We start with the small diagram in Figure 6.1b, which is the supply-demand curve for an individual speculator, and has the same form as D_s-S_s in the large diagram. Defining E as excess demand (= D − S), we see that for the coordinate (E_j',P) we have $P = \overline{P}_j - m_jE_j'$ (where for *this* speculator, at point (E_j',P), we have $E_j > 0$, with the same sign holding for the value of m_j). This expression yields $E_j' = (\overline{P}_j - P)/m_j$, and holds for any 'j': that is, for (E_j',P) we have a demand for futures contracts by *this* speculator; but at the same P we might have a supply by another speculator. (And observe: $P = \overline{P}_j - m_jE_j'$ is merely a linear equation in the (E_j,P_j) plane, with $\overline{P}_j$ the intercept on the vertical axis, and m_j the slope.) We can now look at the excess supply situation for all speculators.

$$\sum_j E_j = \sum_j (\overline{P}_j - P)/m_j = \sum_j (\overline{P}_j - P)m_j' \qquad (m_j, m_j') > 0 \quad (6.1)$$

In this expression $m_j' = 1/m_j$. Assuming that there is some price $\overline{P}_f$ at which the market clears, we get:

$$\sum_j (\overline{P}_j - \overline{P}_f)m_j' = 0 \qquad (6.2)$$

This can be solved for the market clearing price, $\overline{P}_f$.

$$\overline{P}_f = \frac{\sum m_j'\overline{P}_j}{\sum m_j'} \qquad (6.3)$$

This market clearing price for speculators, $\overline{P}_f$, can be seen intersecting the vertical axis in Figure 6.1a. It is the price which causes the excess supply (or excess demand) of speculators to balance in a market where there are *only* speculators. Things change when we add hedgers. As will be shown later, when they come into the picture the market clearing price becomes P_f^*.

Some further elaboration on the expressions 'long' and 'short' might be useful here. A merchant who holds a physical inventory is considered long in the spot market, because that trader is in possession of the commodity. A trader is short in the forward market if the trader has signed a contract to deliver a certain amount of the commodity at a known or unknown price. All the time here we play on the theme of long being associated with buying, and short with selling. Furthermore, traders who are net long or net short in some market are often considered speculators. Of course, considering all 'uncovered' positions speculation might be going too far, since it would mean that if your house, car, and other insurables were not insured down to the last dollar of their worth, you are a speculator.

If a speculator sells a futures contract, then he is short in the futures market. A transactor whose net exposure in the actuals market is offset by her position in the paper market is a hedger: she is a short hedger if short in the spot market and short in the futures market, and a long hedger if this situation is reversed. Remember also that futures markets, when they exist, are centralized and trade only standardized contracts, for specific amounts, with fixed delivery periods and delivery locations, although increasing flexibility is becoming the rule. It is the *forward* market where all sorts of *ad-hoc* variations are possible in regard to price, quantities, delivery conditions, etc.

We can now complete the discussion associated with Figure 6.1a. As compared to speculators, a hedger (in e.g. oil) would not normally buy *and* sell contracts for the same commodity, but open positions in 'paper' oil corresponding to that in 'wet' barrels: buyers of oil buy (go long in) futures contracts, while sellers of oil go short in futures.

Assuming that short hedgers find hedging increasingly attractive the higher the price of futures contracts, their supply curve takes on the appearance shown in Figure 6.2a. Similarly, a long hedger, being afraid of a price rise, might find futures attractive if their price was low. (Pay particular attention to the word 'might'. What it means is that a low futures price might signify to some transactors the likelihood of a low spot price in the future.) Observe that in this exposition a long hedge means positive excess demand for futures. The aggregate supply-demand curve for hedgers is S_h–D_h Figure 6.1a, and the overall futures market equilibrium for that particular item, having that particular maturity, is at point (E^*,P_f^*). (But note, the aggregate $S_h - D_h$ curve may be 'broken' at the vertical axis.)

At the equilibrium shown here, hedgers, in the aggregate, are short (i.e. protecting against price falls), while speculators are net long: on the average, they expect futures prices to rise! Of course, *some* hedgers are net long, just as some speculators are net short. Many observers consider this the normal situation, given the very large inventories of oil that are held. Net long hedging has been observed in the oil market, however, in situations where inventories were regarded as insufficient. An example of this occurred during the Gulf crisis, when inventory holders did not hedge short, but instead bought futures contracts in anticipation of higher oil prices.

6.2 Further Mechanics of Hedging and Speculation

We have now come to a crucial subject: the convergence of the futures and spot price of a given physical commodity. It is only when these converge that a futures market can provide a satisfactory hedging of price risk, and in theory this should always be the case when the seller of a futures has the right to make delivery, and can make delivery, while the buyer of a contract can hold the contract open and so obtain delivery. If we observe real-world situations, we see that if we have a high degree of liquidity in the futures market, and no geographical and technological obstacles to making or taking delivery, then price risk can usually be hedged satisfactorily. Otherwise, hedging can be unsatisfactory, and a great deal of money can be lost.

Conventionally, when a contract reaches the expiry date, the shorts who have not offset their positions must deliver the item in the manner designated by the contract, while the longs with open positions must accept and pay for what they have bought. Thus, at the time of delivery, if the spot price (S) is greater than the futures price (F), the longs will accept delivery and sell on the spot market. This arbitrage should work to equate S and F. Similarly, if F > S, the shorts will buy on the spot market and make delivery (rather than closing their position with a 'buy' at the higher futures price). Once again, arbitrage should move S toward F.

But we can go further. The shorts would not deliver anything that could be sold at a higher price in the spot market, and so, as the delivery date approaches, if S > F they would offset their futures contracts. This amounts to an increase in the demand for futures that raises their price, and thus drives F closer to S. At the same time, spot sales should press S down toward F. Similarly, the longs would not want to take delivery of anything that they could buy more cheaply elsewhere. Thus, if F > S as the delivery date approaches, they will offset their contracts. This amounts to an increase in the supply of futures, which tends to depress F, while their demand for the physical commodity raises its price: S increases. Naturally, in real markets this mechanism does not work perfectly, since in commodity markets the transaction costs associated with the inconvenience of delivery usually prevent a complete closure of the two prices; but about 15 years ago Philip

Verleger presented some research which showed that for several energy items a very satisfactory amount of convergence was experienced.

Although crude oil has been satisfactorily hedged with futures contracts for heating oil, there is no logical reason to expect price convergence for non-homogeneous goods. The same problem would exist if a transactor tried to hedge Brent (North Sea) oil with West Texas Intermediate: the geographical difference between these markets could allow prices to move in opposite directions. The problem of non-converging prices (at the expiry date) is called *basis risk*; and sometimes basis risk is used to describe a diverging spot and futures price during the 'running time' of a futures contract, even though these prices eventually converged. A hedger might be called upon to pay huge margin charges because of such a divergence, even though she expected to recoup her losses eventually – assuming that we ignore the time value of money.

The word margin was used above, and in Chapter 3, but its full significance needs to be grasped. For each contract bought or sold, traders must post a small percentage of the nominal price of the contract with their broker. This amount, which is a kind of deposit, is known as margin. Contracts are then revalued or *marked-to-the-market* after the close of each trading day, usually at the market's closing price. If the closing price has moved below the previous day's closing price, then traders who have bought contracts (i.e. who have long positions) might be asked to pay additional margin, while those who have sold contracts (i.e. have short positions) will show a profit on their contracts that, if they wish, can be taken in cash, or simply kept on deposit with their broker. This process ensures that profits and losses are not carried too far forward.

Futures exchanges, like stock exchanges, are membership organizations, and non-members trade 'through' members, i.e. brokers. In addition, almost all futures exchanges have an associated *clearing house*. This institution acts as seller to buyers, and buyers to sellers – which means that it functions as an intermediary in the transfer of monies between brokers, who in turn are representing their trader clients.

Now we are in a position to understand how a futures market permits us to *lock-in* a price. Assume that in 30 days you must buy 1,000 barrels of oil for your firm. The spot (physical) price today is $19.5/Bbl, and the futures (paper) price is $20/b. You therefore go long for one contract (=1,000 barrels) with a maturity of 30 days, and begin to observe the price of your contract.

Suppose that the price rises by one dollar/day for 30 days, at the end of which time you offset your position by selling a contract – thus closing your position. Every day your contract is marked-to-the-market by one dollar/barrel, and so you earn $30/b by the time you offset your contract.

(Your total gain is $30,000.) But regardless of what happened to the price of physical oil during the month, it should (ideally) converge to the price of the paper oil by the maturity date, which is in 30 days. You must therefore pay $50/b for physical oil, but given your gain of $30/b on the paper oil, the net price is $20/b. You have locked in a price of $20/b. (Note that you did not lock in the spot price of $l9.5/b.)

There is an implication above that after you buy your futures contract, you can forget about what happens in the actuals market. This is not quite true. If, in the example above, the price of physical oil fell rapidly, and you came to give immediate consideration tdo offsetting your futures contract, since a contract losing value is one on which margin must be paid.

The last topic in this section concerns something called the exchange of futures for physicals (EFP), which is no more than an arrangement which permits buyers and sellers of futures contracts to negotiate deliveries under terms which differ from those set out in the contracts they are holding.

The basic EFP is an agreement between two transactors with opposite futures market positions to match both physical and paper commitments, and make and take delivery at a mutually agreed on time: in the case of the EFPs that have been taking place on the New York Mercantile Exchange (NYMEX) for almost a decade, transactors inform *NYMEX* (via their brokers) that they want to offset each other's futures contracts without employing the normal mechanisms of the exchange.

If transactors are holding futures contracts with the same opening date and maturity, then a possible price at which the physical transaction will take place could be the futures price on the day the contracts were opened, although in theory it could be the price on any date that was acceptable by both parties. In addition, if transactors want to deliver a different grade, or take delivery at a time and place that was not specified on the contract, this could also be arranged with an EFP.

NYMEX is one of the three great energy exchanges in the world. The other two are the International Petroleum Exchange (IPE) of London, and the energy section of the Singapore Monetary Exchange. (NYMEX has long had hopes of merging with the IPE.) The IPE is also attempting to introduce more flexibility into its activities. Delivery is not necessary on IPE contracts. Instead, the contract is settled in cash against an index published daily by the IPE. At the maturity of the contract the original price is compared to a settlement price based on the index, with payments being made to contracts showing a profit, at the expense of those showing a loss. It does not have to be said that Singapore will not be far behind where innovations of the above nature are concerned, and will very likely introduce some in its own.

6.3 Basis Risk

Basis risk is an extremely important matter, as a number of senior executives at Metallgesellschaft (MG) found out not too long ago, and they are not alone. Even the United States government's Energy Information Administration (in *Natural Gas Monthly*, March, 1990) has admitted that the substitution of basis risk for price risk – which is often inevitable in futures markets – may result in a failure to reduce financial exposure in natural gas markets.

Some simple algebra should illuminate the difficulty, but first let me point out that I will be using something called the *expectations operator*, which will be designated E. For instance, in thinking about the price of oil on 1 January of next year, I might write $E(p)$ – the expected price; while in thinking about a price of the same commodity on January 1 of last year I might simply write 'p', the *ex-post* price. Let me also use this opportunity to note that the following discussion is extremely important, and readers should adjust their concentration accordingly.

The basis is defined as $S - F$, where S is the spot price, and F is the futures price. If we commence with short hedging, then hedging takes place because of the possibility of a loss due to a fall in S, or $E(S_1) - S_0 < 0$. Here '0' signifies the time at which both paper and physical transactions are opened, and '1' the time at which they are closed. Theoretically, this expected loss would be offset by a profit in the paper market of $F_0 - E(F_1) > 0$. We thus write as our expectation of a net profit, V_s, from a short hedged position, the *ex-ante* (before) relationships:

$$E(V_s) = [E(S_1) - S_0] + [F_0 - E(F_1)] \qquad (6.4)$$

(or)

$$E(V_s) = [E(S_1) - E(F_1)] - [S_0 - F_0] \qquad (6.5)$$

Using the definition of the basis ($B = S - F$), we can write (6.5) as:

$$E(V_s) = E(B_1) - B_0 \qquad (6.6)$$

Turning from an ex-ante to an ex-post scrutiny of this operation, with $E(B_1)$ replaced by B_1, and $E(V_s)$ by V_s, we see that if S and F move in the same direction by the same amount, the profit on the hedged position is zero, and the hedge is said to be 'perfect'. But if the basis changes, and changes in the wrong direction, we could find outself with a negative V_s. Now we should understand why the hedger has substituted a basis risk for the price risk that he or she is trying to avoid.

To see what could go wrong here, take the initial basis $(S_0 - F_0)$ as zero, and for the ex-post (realized) basis take $F_1 > S_1$ (which is a condition known as *contango*). As we see from an ex-post version of (6.5) or (6.6), a *decline in the basis* gives $V_s < 0$. (If $F_1 > S_1$, then $B_1 < 0$, and so (ex-post) $V_s = B_1 - B_0 < 0$, since we had $B_0 = 0$).

Why doesn't arbitrage force the convergence of S_1 and F_1? In an ideal situation it would, but if commodity X is being hedged with commodity Y, or if the commodities are homogeneous but, for example, European oil is being hedged on a US market, etc, then price convergence might not come about. Of course, in our example, if the basis had increased (e.g. to $S_1 > F_1$, which is a condition known as *backwardation* or *inversion*), then a positive profit would have been realized. This suggests that we get the antithesis of the above result if we commence with a long hedge for the purpose of guarding against a price rise. Our ex-ante relationship is:

$$E(V_L)=[S_0 - E(S_1)] + [E(F_1) - F_0] \tag{6.7}$$

The rationale for long hedging is usually that $E(S_1) > S_0$, and so by going long, if the spot price does increase, then $E(F_1) > F_0$. By rewriting (6.7) we get:

$$E(V_L)=[S_0 - F_0] - [E(S_1) - E(F_1)] = B_0 - E(B_1) \tag{6.8}$$

Ex-post, if the basis $(S - F)$ increases, there will be a loss. For what it is worth, MG was involved with a strategy of continually rolling short-dated positions (in both futures and swaps) forward, and unforseen changes in the basis led to a steady stream of ill tidings. (See the appendix to this chapter for the mechanics of rolling a hedge forward.) MG's losses turned out to be catastropic, but there might be some truth to the argument that their misfortune was due more to bad luck than to bad strategy: had the basis changes they experienced been essentially random, then in the long run their losses should have been bounded by transaction costs plus, perhaps, some small risk premiums. Of course, the great unasked question is why MG failed forget about futures and 'try' to concentrate its hedging on swaps.

6.3.1 Exercises

1. Given the situation in Figures 6.1b and (6.2) for individual speculators and hedgers, show how to aggregate supply-demand curves for both categories of transactors. Is there anything peculiar here?

2. Many airlines hedge their jet-fuel requirements. Discuss how they could do this with futures if there was no jet-fuel futures contract. What problems might arise?

3. In Figure 6.1a we have hedgers net short and speculators net long. Shift the speculator   supply-demand curve in such a way as to obtain the opposite situation. Explain how this might come about!

4. An oil refiner is afraid of a price fall for her products, and a rise rise for her most important input. What are some of the products that I am talking about, and what input? How can she hedge?

5. Why bother with derivatives markets. Why not just hedge with forward contracts?

6. How can you speculate using forward contracts?

7. What does backwardation mean for you if you were hedging long, and is there a basis risk?

8. Suppose that you have a nice portfolio of 10 different shares, and you decide to hedge in futures using the Standard and Poors (S&P) futures contract consisting of a bundle of 100 different shares – suitably weighed, of course. What problem might you have?

9. (Extra credit) The ex-post version of equation (6.8) is $V_L = B_0 - B_1$. We can thus write $V_L - E(V_L) = E(B_1) - B_1$. Since stability of the basis is a key to efficient hedging, we might expect that things will go well for us since $E(B_1) = B_0$. What happens if our expectations are wrong, and it turns out that $B_1 > B_0$? $B_1 < B_0$?

10. In the last paragraph of the text I use the expression 'bounded by'. What does this mean? What does 'risk premium' mean?

6.4 An Options Fable

Wedding bells will soon be ringing. The socially prominent Ms Evelina Grundy of Aberdeen (Washington) and Algiers (Louisiana) has announced her engagement to former US Marines Sergeant Hoover Lee Clarke, a popular Roper County (Alabama) sharecropper. There are, however, a few problems.

The main problem is money for the wedding and the wedding reception. Things have not gone well for Hoover since the man who pulled his plow departed to take a job at Hamburger Heaven; and while the Grundy family fortune was founded during the Civil War on houses and lots, later generations of Grundy men dissipated it in the same manner. The truth is that these two delightful people must throw themselves on the tender mercies of Hoover's oldest son, Hector Lee Clarke. Once a teacher of nuclear physics, Hector is now a quantitative analyst (or 'quant' or 'rocket scientist', as they are sometimes called) at Gekko and Montana, a well known Wall Street (N.Y.) investment bank. Hector accepts the assignment with great reluctance; but after a few days of thinking about it, he and his wife, Lucy Sweet Clarke, become enthusiastic. Hector's father, and his other male relatives, have always considered him a blight on the family name, since most of them were moonshiners, pool sharks, 'do-ah' choristers, stand-over men, and hoboes. Now he and Lucy have a chance to show these 'good-old boys' what the good life in a 12 room duplex on Park Avenue is all about.

The wedding will be on a Friday, and the reception is in the form of an open house lasting from that evening until midnight on Sunday. Unfortunately, there will be hundreds – if not thousands – of guests, most of them invited to the affair by Evelina's seven brothers. Food is going to be a problem, but not nearly as much of a problem as paying for it: Hector Lee told his employers that the Japanese share market would escalate during the last few months of 1996, and instead, during the first week of 1997, it came close to a 'melt-down'. Mr. Montana then gave Hector an offer he could not refuse: if Hector forgot about his bonus and his annual raise in salary, then he might escape the notoriety associated with 'security' escorting him to the front door, and dumping him and his personal effects into the gutter. Hector's memory, which he always boasted of as being perfect, suddenly became extremely bad, although he did remember to come to work earlier every day, and stay later.

The cuisine will consist of finger-food, ordered from the best caterers. Huge amounts of it. Evelina and Lucy Sweet Clarke decided on this because, when Hoover Lee is in his cups, he eats everything with his fingers except his soup: his soup he drinks directly from the bowl, and so there will not be any soup. When planning for the party begins, finger-food is selling for $40/kilogram, but Lucy has no idea at all of how expensive it will be on April 1, when the reception commences. If only Hoover Lee had married a Grimaldi or someone affiliated with the Court of St James. But, as Hector's brother Harry was fond of saying, "why be in a war without learning something from it". The war that she and Hector had been a part of for the past ten years was the war fought in trading rooms and offices, and the corridors and restaurants of power, in New York and the City Of London, Frankfurt and Tokyo. It was a war to which only people who knew how to make large amounts of money, and keep it, were invited. She picks up the phone and calls her broker, Minnie Palmer. "I'm going to make this thing work", she said aloud.

Lucy purchased some options contracts which gave her the right to buy 1,000 kilograms of finger-food by April 1 for $50/kilogram. This kind of contract is called a *call option*, where the $50/kilogram is called the *striking* (or *strike*) or *exercise* price. When the initial transaction takes place, Lucy has to pay a *premium* to the seller of the option, who is also called the *writer*. For the purpose of this exercise, let us take this premium as $5/kilogram. This money (= $5,000) is transferred immediately to the option writer via Ms Palmer.

Notice the position of the writer of the option. On or before April 1, he must be prepared to deliver 1,000 kilograms of finger-food to Lucy for $50/kilogram. If the market price at which he purchases finger-food – in the event Lucy exercises her contract – is $100/kilogram, or for that matter $1,000, he must still obtain these victuals. Thus, theoretically, his losses are unlimited. But on the other hand, if the market price is less than $50/kilogram when Lucy orders her finger-food (e.g. $44/kilogram), then the option writer simply breathes a sigh of relief, since he is $5,000 (= 5 x 1000) to the good. In the latter circumstances, Lucy is also happy. She was able to afford 1,000 kilograms of finger-food at $55/kilogram (which was the premium + the striking price), but with the market price at $44, she does not need to exercise the option (which would have meant buying the finger-food for $50/kilogram). Instead she purchases her supplies for 1,000 x 44 = 44,000 dollars, and gives a marvelous reception. (She also forfeits the premium, of course.) Something else that needs to be observed here is that unlike a forward contract (which involves a firm commitment to buy or sell something at a future date), or a futures contract, the option contract need not be exercised.

That immediately raises a question: why didn't Lucy use a forward or a futures contract?

Had she approached her caterer about a forward contract, he might have told her that with all the parties and receptions taking place around the first of April, the best he could do would be to offer finger-food at $70/kilogram, which Lucy considered too expensive. But even if he had said $50 or $55, she might have come to the conclusion that it was best to pay a relatively small premium to keep her options open. This decision apparently paid off. As for a futures contract, it might have permitted her to lock-in a finger-food price close to $40/kilogram, but unfortunately there was no futures contract for finger-food. Ms Palmer suggested trying to hedge using a contract for Long Island Trout, which are also consumed in large amounts about that time, but warned Lucy that the basis risk might be considerable. Having read the previous section of this book, Lucy abstains.

The reader should also take note of when the option contract is exercised. Just as a viable futures contract can be offset at any time, an American option can be exercised at any time. On the other hand, a European option can only be exercised on (or perhaps around) the *expiry*, or *maturity*, date. Most options sold in the world are American type options, to include those sold in Europe.

Next we consider the important matter of how the exercise price and the premium are determined. Suppose that on February 1, when all decisions about the reception have to be made, finger-food is selling for $38/kilogram. It might thus be suspected by option writers that an increase to $50 dollars by April 1 is unlikely, but far from impossible. Options are often traded over-the-counter (OTC) rather than on an exchange, and so when Ms. Palmer checks the premiums for 3-month finger-food options on her computer, she sees the following for 3 month call options

Finger-food: Exercise Price,	Premium
40	10.35
45	7.45
50	5.00
55	4.15
60	2.55
65	1.33

When we begin to put some of this in algebraic form, the premium will be referred to as the price. A point of considerable importance is that if it were commonly thought that there would be a large amount of finger-food available around the first of April, then the chances are that the market price of this item should fall, and very few party and reception givers would be willing to pay the prices shown in this tableau. This would mean that prices (i.e. premiums) would have to decrease also, everything else remaining the same.

But everything else may not remain the same. Many buyers of finger-food call options are not party givers like Lucy, but speculators. These are people who are betting that the price of finger-food will rise, if only momentarily, by a large amount sometime before the maturity date of the contract. If this happens they will exercise their options, and sell the finger-food on the spot market. For instance, suppose that for some strange reason the price of finger-food jumped to $100/kilogram on February 20. The holder of a call option with an expiry date of April 1, which was purchased for a premium of $5/kilogram, with a strike price of 50, might then exercise the option. This means that she purchases finger-food from the option writer for $50/kilogram, who in turn has to purchase it on the spot market for $100/kilogram. The option buyer might then immediately sell it on the spot market, realizing a profit of $100–50–5 = $45/kilogram.

Thus, even if an ample supply of a commodity is expected, there might be a price 'spike' due to a temporary surge in demand, or a sudden reversal in expectations, and even though there might be a return to normality in a few days, the options writer has still lost a great deal of money. Consequently, in determining premiums, the length of time to the maturity date, as well as the price volatility of the underlying asset or commodity, is extremely important. Even if a bumper supply was expected, an option contract might still have a high price if the expiration date of the contract was in the distant future, and historically finger-food prices were extremely volatile.

The value of the option premium that is exclusively associated with the length of time to the expiry date, is called the *time value* of the option. Thus, in a secondary market for options, an option that is sold only a few days before its expiry date would have hardly any time value left. According to routine theory, supply and demand determine premiums, and writers select a premium-maturity combination hoping that the option will never be exercised, but will simply provide what is known as premium income. Naturally, an ideal options market is open to all comers, and had Lucy decided that she wanted to write instead of buy a e.g. call option, that opportunity should also have been open to her.

An important concept that needs to be considered now is the *intrinsic value* of an option. This is the difference between the actual market price, and the exercise price of the option. If the exercise price was 50 dollars, and the market price $58/kilogram, then the intrinsic value of the option would be $8. Furthermore, this option would be *in-the-money*, because it would make sense for someone who paid e.g. a $5 premium for the option to exercise it, since the gain would be 58-50-5 = $3/kilogram. On the other hand, had the market price been $43/kilogram, then the option would have been *out-of-the money*, as the reader can show. Note also that an option would not necessarily be exercised because it moved into the money. It's owner might wait, hoping for it to move deeper into the money.

In this informal introductory discussion, only call options were considered. But we might easily have a situation where a seller of finger-food would like to ensure that he would not have to sell for less than $50/kilogram. That gentleman would then buy a *put option*, which means that if he exercised it, the writer of the option would be obliged to buy finger-food for $50, regardless of the actual market price. As in the previous example, the writer of the option fervently hopes that the option will never be exercised, and hopes that the premiums are set in such a way as to make this wish come true. (Although they should not be set so high as not to tempt buyers.) Note also that in this situation a writer's losses are limited, since the market price cannot fall lower than zero. For example, if a put option with a 5 dollar premium and a 50 dollar exercise price were sold, and the market price fell to 25 dollars, then the buyer of a put option could buy finger-food on the market for $25/kilogram, and sell it to the option writer for $50. The buyer's profit would be 50–25–5 = $20, which is the loss of the option writer. In this situation the intrinsic value of the option is 50–25 = 25, and clearly it is well in the money.

That brings us to the end of this informal review of options fundamentals. Something that the reader should notice is that in most respects options are easier to understand than futures. For instance, it is conceivable that Lucy could have bought a call option from her caterer, simply paying this person $5/kilogram for the right to buy finger-food on April 1 for $50/kilogram. No formal options market was necessary

6.4.1 Exercises

1. How can you hedge with commodity call options?
2. How can you speculate with commodity call options?
3. How can you hedge with commodity put options?
4. How can you speculate with commodity put options?
5. In the example with put options above, it was said that the writer's losses were limited. What about with call options?
6. What is the difference between 'intrinsic value' and 'time value'?

6.5   The Simple Algebra of Options

Suppose we forget about the imminent Clarke nuptials, while hoping of course that Hoover Lee does not lose his cool at the reception being so generously arranged by his son and daughter-in-law, and does or says something that will leave a bad taste in the mouths of those who are near and dear to him. Instead, let us attempt to turn the previous discussion into something for readers who are afraid that with the arrival of the 'information society', economics and finance will be reduced to the level of video clips. Once again notice will be served that we are not moving beyond secondary school algebra.

We already know that if you are a speculator and you buy a call option, and prices go up, you could make money; while if you write a call option and prices went up, you could lose money. Similarly, if you bought a put option, and prices fell, then you could make money; while if you wrote a put option and prices fell, you could lose money.

We are aware that if you want to hedge against a price rise, you buy call options; while if you want to hedge against a price fall, you buy put options. We also know that options do not have to be exercised: if we are on the buy side we can simply forget about them, although we lose the premium. Option writers hope that this will happen: the option will never be exercised, and they can walk away with the premium. This also implies why many transactors choose futures markets when options markets for the same asset or commodities are available: they dislike starting their adventure by paying a premium. Clearly, however, options display a highly desirable flexibility when compared to futures. With futures the hedger thinks in terms of locking in a price and, *ceteris paribus*, not making a profit (since the underlying logic is that what is gained on the paper market is lost on the 'actuals' market, and vice-versa). With options, favorable price movements can be taken advantage of to yield large savings (as in the case of Lucy in the previous section), or – depending on the nature of the transaction – profits.

Now let us forget about finger-food and go to oil, where producers of oil are logically interested in hedging against price falls, and e.g. refiners are afraid of price rises. If we start out with *put* options, then the expressions for the profit of buyers and sellers are given in (6.7) and (6.8). The symbols employed are S for the price of physical oil, P_p for the price (i.e. premium) of a put option, and E the exercise (or strike) price. (*Note: this E is not the E used in Section 6.3!*) The put option profit expressions cover only one unit of the underlying commodity (i.e. oil).

$$\text{Buyer's Profit} = V_B = \text{Max}\{(E-S-P_p), -P_p\} \tag{6.7}$$

$$\text{Seller's Profit} = V_S = \text{Min}\{(P_p + S - E), P_p\} \tag{6.8}$$

The interpretation of these expressions is simple, although before an example is given for buyers of options, the reader should notice that we have a zero-sum type of situation here, with the gain of one transactor being the loss of the other. Suppose, for example, that S is \$15/b, while the exercise price is \$30/b, and the premium is \$5/b. Then the buyer's profit is calculated from $V_B = \text{Max}\{(30 - 15 - 5), -5\} = 10$. On the other hand, suppose that we have S = 50. Then $V_B = \text{Max}\{(30 - 50 - 5), -5\} = -5$. In other words in the latter case, the buyer would not exercise the option. We know without calculating that the profit of the seller in these two transactions is −10 and 5, respectively, but it might be useful to check these results. In the first case $V_S = \text{Min}\{(5 + 15 - 30), 5\} = -10$, and in the second we have $V_S = \text{Min}\{(5 + 50 - 30), 5\} = 5$.

Figure 6.3 illustrates the situation for put options. In Figure 6.3b the diagrams are drawn employing numerical values from the first example in the previous paragraph.

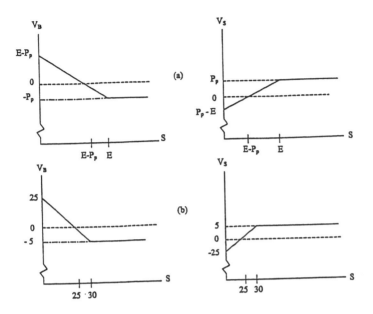

Figure 6.3

Up to now the function of the option writer appears to be limited to absorbing large losses, with only the prospect of an occasional small gain to give him comfort. This is apparently even more true with a call option, since in theory the possible losses of call option writers are unlimited. (The maximum loss of the writer of a put option is $(E - P_p)$, as can be seen from Figure 6.3a.) Still, as will be pointed out later, it can make sense to write options.

Now let us briefly consider call options. Taking P_c as the price (premium) of a call option, we can write for the profits of buyers and sellers:

$$V_B = \text{Max}\{(S-E-P_c), -P_c\} \tag{6.9}$$

$$V_S = \text{Min}\{(P_c+E-S), P_c\} \tag{6.10}$$

The profit diagrams for a call option are shown in Figure 6.4

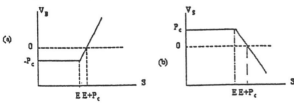

Figure 6.4

As mentioned earlier, potential profit is theoretically unlimited for the purchaser of a call option, just as potential loss is theoretically unlimited for its writer. This can be seen either from the above figure, or from the manner in which S enters the expressions for V_B and V_S.

Next we can look at the algebra associated with the intrinsic value of an option. With a put option the intrinsic value (I_p) can be written $I_p = E - S$. If $E > S$, then $I_p > 0$, and the option is said to be in-the-money. By way of contrast, if $S > E$, the put option is out-of-the-money. In addition, trivially, if $S = E$, then the option is at-the-money. Conventionally, the intrinsic value of an out-of-the-money option is set equal to zero, and so $I_p = \text{Max}(E - S, 0)$.

Now for a very important point, at the time the option is written (i.e. initially sold), the option price P_p cannot be less than its intrinsic value. If it were less, then it would be possible to buy the option, exercise immediately, and realize a profit. Put another way, if we have a profit then $E - S - P_p > 0$, which implies tht $P_p < E - S$: if such an option were mistakenly written, arbitrage would soon eliminate profits. If we consider a situation where $S = 20$ and $E = 22$, then the contract is in the money – i.e. it has intrinsic value. It will therefore trade for at least 2 dollars (or, realistically, more than 2 dollars).

For some readers, a slightly different approach might be useful. Suppose tht on the day the option is traded we have $E - S > P_p$. This gives us immediately $E - S - P_p > 0$. But from V_B and V_S for put options we obtain $V_B = \text{Max}\{(E - S - P_p), -P_p\} > 0$, and with the same kind of manipulation $V_S < 0$. Accordingly, no rational person would write this option, which allows us to deduce immediately that we must have $P_p \geq E - S$. Figure 6.5 summarizes these results for put and call options.

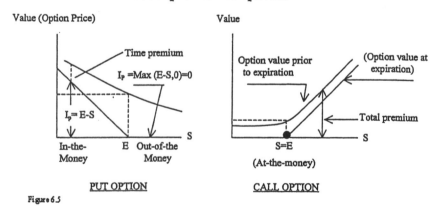

Figure 6.5

In the above diagram it is hardly possible to avoid noticing the importance of the *time value*, or *time premium* as it is sometimes called.

Premiums for options that are out-of-the-money reflect this time value. In-the-money options have time value *and* intrinsic value, which on occasion can make them very expensive. As the reader can see from Figure 6.5, time value is greatest when the option is at-the-money, because in that situation the option is presumed to have a 50-50 (i.e. even) chance of moving into-the-money. As a result the option price should be such that if it did slide into-the-money, it would not be worthwhile for the buyer to exercise it immediately: a profit would be obtainable only if it moved deep-into-the-money.

For a call option having a premium P_c, we have for the option's intrinsic value $I_c = Max(S - E, 0)$, with $S - E > 0$ being defined as in-the-money. In addition, as in the previous case with a put option, we expect the option to sell for more than its intrinsic value, or $P_c > S - E$. Otherwise we have what amounts to a gift by the option writer to the option buyer of $S - E - P_c$, which is greater than zero.

This section will be closed with a few abstract comments, since it appears that option writers accept huge risks. Generally speaking, the option writer is accepting a small probability of a large loss in return for a large probability of a small gain. On the other hand, the option buyer exchanges a large probability of a small loss for a small probability of a large gain. Readers with a background in statistical theory might therefore jump to the conclusion that neither buyers nor sellers, taken as a group, will gain, although individual transactors who are in the market for a short time might do well or poorly; but it appears that, as a group, speculators in options tend to make money in the long run. This is because the risks that are associated with price uncertainty, and in particular price uncertainty brought about by a high degree of volatility – and which occasionally menace the financial survival of hedgers – results in the carrier of these risks obtaining a price for his or her services in excess of that which would be obtained in the utopian setting of a journal article, with its auction market (and perfect competition) format. Real world option markets are usually of the OTC type, where it is not unknown for owners and some employees to make great deal of money, and where casual empiricism indicates that at least a few alert speculators can do well. The same is true for futures.

Commodity derivatives markets, like financial markets, seem to be moving in the direction of electronic, around-the-clock, around-the-world, buying and selling. Some observers claim that the greatly increased market surveillance will result in a higher degree of price instability. Invoking the great German physicist Werner Heisenberg's *uncertainty principle*, the highly successful trader Bruce Kovner suggests that the closer price patterns are observed on financial markets, the greater the likelihood that false signals will be generated. We had an example of this sort of thing during the

158

run-up to the Gulf War, although the exact problem in that situation was
market actors allowing their imagination to overwhelm their capacity for
logical thinking. Even more inexplicable was the movement of the oil price
following the opening of hostilities.

After a few simple exercises there will be a brief discussion of the option
price, to include a presentation of the famous Black-Scholes formula. This
expression was labeled by *The Economist* (February 16, 1991) the most
useful contrivance in "all" economics, which was probably the overstatement
of the year. However there is no denying the logical appeal of this formula in
the light of the above discussion, and there are many observers who
considered the late Fisher Black the leading financial economist in the world
(although my nomination for that title is Merton Miller). The Black-Scholes
formula recognizes time value by including S and E, and the time to expiry
(T). There is also an explicit recognition of volatility, via the standard
deviation of the historical series for S. Here the reader should recall that if S
is highly volatile, then it might move, if only temporarily, into the money –
or deeper into the money; and if the option is an American type option, it
could be exercised immediately, in which case its writer could lose a great
deal of money.

6.5.1 Exercises

1. Make a numerical example to go with Figure 6.3, and calculate V_S and
   V_B.
2. Show *formally* - i.e. without any numbers - that equations (6.7) and (6.8)
   represent a zero-sum arrangement.
3. Consider two options that are identical with regard to exercise price,
   maturity, etc, except that one is an American option, and the other a
   European option. Which should have the highest price?
4. In a formula for the option price, what part should S and E play?

6.6 A Comment on the Option Price

We know from the previous two sections which factors influence the
price (i.e. the premium) of an option. What we really desire is a formula
which tells us how these factors interact, and which provides us with a
'number' that will allow us to make some money in a highly competitive
market. In her extremely informative book 'Trading Up' (1989), Nancy
Goldstone said that the Black-Scholes formula, or at least some variant of
this formula, was in use at the investment bank where she was an executive
in the options trading division.

Strictly speaking, the Black-Scholes formula is used to price European call options on nondividend-paying shares/stocks, but it has been said that even a run-of-the-mill rocket scientist can adjust it to deal with other cases – which may or may not be true. Furthermore, the simple manipulation known as 'put-call parity' will provide an algebraic relationship between the price of a call option and the price of a put option (on the same commodity or asset). Now, taking P as the premium, we can examine the following sketch.

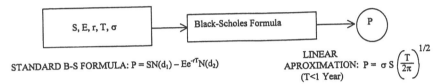

STANDARD B-S FORMULA: $P = SN(d_1) - Ee^{-rT}N(d_2)$

LINEAR APROXIMATION: $P = \sigma S \left(\dfrac{T}{2\pi}\right)^{1/2}$
(T<1 Year)

SYMBOLS

σ : instantaneous standard deviation – a surrogate for volatility

$$d_1 = \frac{Ln(S/E) + rT}{\sigma T^{1/2}} + \frac{1}{2}\sigma T^{1/2} \qquad \text{(and)} \quad d_2 = d_1 - \sigma T^{1/2}$$

$N(d_1)$ = cumulative probability distribution for the standard normal variate, from $-\infty$ to d_1. $N(d_2)$ = the same for d_2.

T: Time to expiry in fractions of a year $(0 \leq T \leq 1)$

r: riskless rate of interest

Figure 6.6

If the reader has some problem with this presentation, do not worry. The point is that, given numerical values for the items in the left hand box, the Black-Scholes Formula, or one of its facsimiles, can produce a value of P. Personally, I find it difficult to believe that this P is the one I really and truly want – the one that will make me a millionaire. Instead, I think it is the one that I can use in a seminar in Stockholm or Sydney, assuming that I use it before my audience begins to yawn. (Note the simple approximation given in the figure when T < 1 year.)

No more will be said about the Black-Scholes Formula, except that it is one of those things that all economics students should have at least a passing acquaintance with. I would, however, like to amplify a procedure shown in many elementary finance books to capture the logic behind option pricing, and which does not require any complex mathematics. The analysis turns on the construction of a synthetic option, and it may provide an opportunity to see how the interest rate enters the picture.

Suppose that an item's price can go up to U, or down to D over a specified period of time. (In case the reader wants to construct a numerical example, the best interval is probably a year.) Assuming that the initial price of the commodity or asset is S, the rate of interest is r, and a transactor can buy a call option for that period with an exercise price of E, what is the 'fair' price of a call option on that item? With E 'between' U and D, and G signifying 'gain', we have the arrangement in Figure 6.7 for the option:

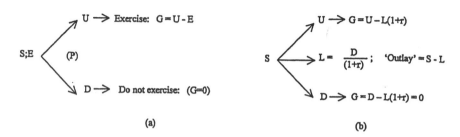

(a)   (b)

Figure 6.7

Note that the gain (G) in Figure 6.7a is the same as the intrinsic value. As for the construction of a synthetic option, this involves no more than replicating the payoffs from the 'genuine' option. These payoffs are $(U - E)$ if the option is exercised, and zero if it is not.

To start we assume that the transactor buys the item for S, while borrowing L. This means that at the end of the year it will be necessary to repay the loan with interest, and thus $L(1+r)$ must be available. The trick now is understanding that the no-exercise arrangement (G=O) involves borrowing just enough so that if the price of the underlying item falls to D, this D will repay the loan. Thus we see from Figure 6.7b that $L = D/(1+r)$. But if the price rises to U, then the gain is $U - L(1+r)$.

Now we are at the point where we can ask how to make the synthetic option that we are trying to construct equal to a 'genuine' option with an exercise price of E. The answer is to have U–E equal to U–L(1+r); and although this is unlikely for a single call option, it would be true for N options if:

$$N = \frac{U - L(1+r)}{U - E} \tag{6.11}$$

A misgiving could arise here if N were not an integer, but luckily this is not a problem because when calculating the price (or premium) P, the configuration of N is unimportant. Now we recognize that the 'outlay' associated with the transaction was $S - L$, and thus a single call option must be worth:

$$P = \frac{S-L}{N} = \frac{(S-L)(U-E)}{U-L(1+r)} = \frac{S(1+r)-D}{1+r}\left(\frac{U-E}{U-D}\right) \qquad (6.12)$$

The above discussion is a long way from Black-Scholes (B-S) country, but it does highlight the important role of prices: if, for example, S falls, then P also declines, since the liklihood of the option being exercised decreases. This suggests that time belongs in the picture – which it is, but not in a really meaningful way. Furthermore, the absence of volatility in equation (6.12) vitiates its applicability to real markets, since volatility is the main determinant of the option price. Finally, this approach does not clarify why – in the present analysis, or the B-S exposition – the (statistical) probability of the asset price rising or falling is apparently irrelevant, although this result is highly counter-intuitive. The answer here turns on the asymmetric nature of an option: the downside is limited to the premium, while the potential profit depends on *how far* the price might move, and the volatility – which is in the B-S equation, though not (6.12) – tells us that. And how, *in theory*, is the B-S relationship used? The answer is that P is calculated, and compared to the existing (market) P. If they are not the same, then arbitrage is *theoretically* possible. Notice the emphasis: *in theory*, and *theoretically!*

6.7 Oil Swaps

Oil swaps are different, and this seems to have limited their popularity in 'academic' circles. But they are very important. Theoretically, there is a way to tie futures and options together, while swaps are an outsider in this context. Fortunately, this makes them easier to understand, and this is true even though swaps can be combined with options in so-called 'swaptions'.

One reason for emphasizing the importance of swaps is that Metallgesellschaft's much talked about derivatives position of 160 million barrels (= 160 MBbl) of oil and oil products was heavily weighted in over-the-counter swaps. The provocative thing here, however, is that these were short-dated swaps, although we know that swaps are becoming increasingly important relative to futures and options because they can involve very long maturitites: fifteen months might be regarded as a long dated commodity futures contract, while swaps of well over 5 years have been recorded. I was also surprised to hear that Mr. Michael Hampton (of Chase Investment Bank) has publicly proclaimed that after 6 months, swaps *are* the derivatives market. This contention may be close to the truth, but it is not very popular because of the absence of price visibility in the swaps market. This absence is one of the factors making a comprehensive theoretical discussion of the swaps market extremely difficult.

As many readers of this book already know, some of us who teach undergraduate finance enjoy attempting to convince our students of the relative simplicity of various derivatives. I have always tried to avoid doing this with certain financial futures, but on a number of occasions I have definitely oversimplified the situation with commodity futures and swaps. Fortunately, the well advertised MG debacle made it clear to me (and I hope others) that serious mistakes can be made with these instruments due to the presence of such things as basis and 'rollover' risk. (Briefly, rollover risk involves having to pay a much higher price to renew your position, which is a kind of insecurity that is often associated with short-term loans. A short comment on rollover risk can be found in the appendix to this chapter.)

Many observers now appear sympathetic to this position, but some of them also claim that if MG's board and creditors had allowed MG's management to ride out the storm, all would have gone well. Instead, the theory goes, they panicked, and brought in an outside team whose clumsy trading inflated some of the mistakes that were made earlier. As it happened though, riding out the storm would have required paying some huge margin calls, and MG's potential creditors reasoned that this money might have been better used.

No attempt will be made in the present contribution to resolve this quandary, although it brings to mind an important consideration that seems to have been completely overlooked. It is constantly claimed that derivatives markets are of crucial importance because they increase the 'visibility' of prices, but if it turns out that in a particular market futures are unsatisfactory for hedging risk in all except very short-run situations, then more emphasis will be placed on swaps – where prices are generally *not* visible. Furthermore, as we know from the new rhetoric of deregulation, the presence of properly functioning futures markets is touted as a *sine qua nom* for viable competition in a world of uncertainty – i.e. a world where long-term (contractual) transactions are played down in favor of spot deals. As a result, certain 'researchers' are making it their business to insist that derivatives markets will always perform satisfactorily, even in those situations where their accomplishments are conspicuously unsatisfactory. (For example, as is well known just now, electricity derivatives in Norway, and natural gas futures in the United States.)

The first thing to appreciate here is that the logic and mechanics of swaps are far simpler than the rituals associated with exchange traded futures and options, since in many respects swaps are analogous to conventional insurance schemes. (The swaps market could also be regarded as a kind of futures market without speculators.)

Specifically, swaps are a not-so-distinct category of what is know as 'structured derivative products'. Other items in this collection include price protection programs, price collars, price caps, corridors, and as we will consider below, so-called participating swaps. All these products are traded on an OTC basis by various financial institutions, instead of via organized exchanges, with these banks, brokerages, trading houses, etc functioning as intermediaries between transactors whose hedging interests are diametrically opposite. In the case of MG, whose total swap position seemed to average about 100-110 million barrels (Mb) of oil and oil products, protection was mostly desired against upward price movements.

This is so because MG held large numbers of forward contracts which exposed it to rising prices, having agreed to purchase a considerable amount of refined products from Castle Energy at a price which locked in a guaranteed profit margin for Castle. Similarly, refined products were being delivered on fixed price forward contracts that would have resulted in substantial profits had MG's expectations of stable or declining crude oil prices been correct.

Something that needs to be clarified here is that the MG we are talking about in the present discussion is the U.S. subsidiary of Metallgesellschaft A.G., Germany's 14th largest industrial firm. This subsidiary was primarily engaged in oil and oil product trading, although it had refining ambitions, and held a large equity interest in Castle Energy which, though originally an oil exploration firm, was helped by MG to become a refiner. Hedging strategy has formulated in Germany.

Now for an example in which the entities involved will be called B, G, and A. In order to hedge against price rises, B enters into an OTC swap in which it receives payments based on floating (e.g. spot) prices, while making fixed payments. Somewhat ambiguously this is called 'paying fixed and receiving floating', assuming that the difference between floating and fixed is positive. Otherwise, instead of receiving, B would pay the difference. In both cases, however, the actual cash flow is a difference: e.g. the fixed price minus the floating price. For this reason, a swap is sometimes called 'a contract for differences'. It also needs to be understood that 'floating' is used instead of 'spot' because this floating price is often an *index* that is constructed from the spot price. For example, an average of spot prices over a month, or over 3 months.

The payments mentioned above are usually made by B to a large swap dealer, which will be labeled G (for 'go-between'). G, which is often a major financial institution, is sometimes designated the counterparty, but the correct designation is *swap arranger* or *intermediary*. In this example the counterparty will be called A, and this entity will be 'receiving fixed while paying floating'. In other words, they will be ensuring that the price they receive for producing the item in question does not fall below a certain level: if it does fall below this level, they will receive 'fixed minus floating' from G, while if it is above, they will pay 'floating minus fixed' to G. (Figure 6.8 and Table 6.1 below should clarify this.)

Now for two caveats. The intermediary, G, could elect not to find a counterparty, for the simple reason that it is satisfied with the profit potential associated with *not* finding a counterparty. Here G is functioning as a kind of option writer, betting that the option will not be exercised, and at the same time always being on the receiving end of a cash flow. For instance, in the case of B above, they speculate that the price will remain low most of the time, and so at the end of each payment period they will receive 'fixed minus floating' from B. Another possibility is that they delay finding a counterparty, or they find more than one, etc. It has been said that the most aggressive intermediaries will always accept being the immediate counterparty, although they may have to be compensated (in one way or another) for doing so. ('Immediate' here means that they will not delay accepting a deal with B or A because they do not have a counterparty.)

In the argot of the swaps business, these self-assertive swap arrangers do not insist on running 'matched books', but will accept large exposures in the interval between accepting to become an intermediary, and finding counterparties. Of course, one of the reasons they will do this is that they are experts in 'laying off' any undesirable risk in futures and/or options markets

In addition, it would be theoretically possible for B and A to bypass G, but while these transactors may have opposite hedging requirements, they may not match exactly with regard to timing, amounts and maturities, etc. Accordingly, it is more comfortable to turn any matching difficulties over to an intermediary. There is also a possibility, as with the clearing house in a futures market, that G has a higher credit ranking than the swap participants, and so B and A do not have to be concerned with annoyances like credit risk.

Continuing with this example, let us assume that B wants a deal involving fixed payments every quarter for 2 years of $20/barrel, while receiving a spot related price. The *notional* amount – i.e. the number of barrels to which payments will apply – is 1000 barrels. (As in financial transactions, the term notional refers – in one sense or another – to the principal, and not payments.) Note also the expression 'spot related'. If payments were being made every day, then the spot related price would likely be the spot price itself, assuming that these prices relate to an identical commodity. Otherwise we need some kind of index to represent the floating price, which takes into consideration both the payment period, and any ambiguities associated with e.g. the commodity description. For example, as pointed out earlier, the floating price might be a simple average of daily spot prices over each quarter. The following diagram summarizes some of this information, since it shows swap parties, the intermediary, and cash flows.

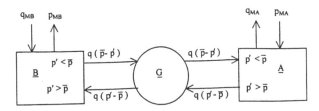

Figure 6.8

In this diagram p_M and q_M signify market prices and quantities, however it is not necessary that $p_{MA} = p_{MB}$, nor $q_{MA} = q_{MB}$. $\bar{p}$ is the fixed payment which, in this example, is made by B and received by A. (As will be explained later, however, neither the fixed price nor the floating index needs to be the same for B and A, since these counterparties are dealt with independently by G.) The (floating) index is, of course, p', and q is the notional amount. Naturally q is not necessarily equal to q_{MB} or q_{MA}, and a similar statement can be made about p' and the market price. As indicated above, in the present discussion both B and A have 'locked in' a price of $\bar{p}$.

Thus far, our example represents an ideal situation: the intermediate (G) simply brings B and A together, receives payments from one counterparty, while making payments to the other (if $\bar{p} \neq p'$) at the end of e.g. every quarter over a two year period. In return for these services G might receive a fee (i.e. commission) every quarter, but often intermediaries such as banks and other financial establishments make their money through price spreads.

In order to understand what this means, in 1991 the average price of oil was $23.23/Bbl, and the state of Texas (US) could have signed a swap contract with First Chicago covering the 18 month period March 1992 to September 1993 for a price of $21.05/Bbl. In other words, $\bar{p}$ was $21.05, and Texas was receiving fixed while paying floating. This was apparently a very attractive proposition for First Chicago, even if an intermediary could not be found. What about Texas? As things turned out, they would have gained from this transaction, since for the period mentioned, the oil price index suddenly fell, and averaged $20.20/b. Thus, had they accepted the arrangement, a deal involving a notional amount of 100,000 b/month (or 100 contracts), would have meant a gain of about (21.05 – 20.20) x 1,800,000 (b) = $1,530,000. Had First Chicago not found a counterparty, which was unlikely, and in addition failed to reinsure via futures and/or options, that amount would have been their loss. There is an obvious lesson to be drawn from this (real life) example. A big reason why oil swaps are so popular might be due to the volatility of oil prices. Otherwise, a producer might be inclined to laugh at the offer of a 'floor' of $21.05 when the prevailing price was $23.23. Incidentally, these figures imply an *ex-ante* price spread of $2.18/b.

In the numerical continuation of the first example above, let us assume that at the end of the first period (e.g. quarter), with the spot price index p' equal to 16, and the fixed unit payment $\bar{p}$ at 20, B pays the intermediary 20,000 dollars while receiving 16,000. The actual cash flow (B →G) is thus 4,000: ($\bar{p}$ – p')q = (20 – 16) x 1000 = 4,000. Put another way, the net transfer for B is –4,000 dollars. For A the situation is the opposite (i.e. +4,000 dollars). Some observers might prefer to view this arrangement as B → G →A. Similarly, at the end of the following period, with p' = 22, B pays G 20,000, while receiving 22,000. The net transfer (A→G→B) is thus +2,000 for B. Observe what this means in the total picture: B buys the *swap amount* in a 'fictional' market for $22,000, and the $2,000 cash flow is a compensation for the amount it must pay over 20,000, which it has locked in with the swap. (In the real market, however, B's outlay is $p_{MB}q_{MB}$, and this does *not* have to be equal to the swap amount. In tabular form, these and the following six transactions are shown in Table 6.1, which represents in a simplified form the payments scheme for a hypothetical swap.

Table 6.1

Period	$\bar{p}$	A's receipt B's payment	p'	B's receipt A's payment	Actual Cashflows	
					B	A
1	20	20,000	16	16,000	-4,000	+4,000
2	20	20,000	22	22,000	+2,000	-2,000
3	20	20,000	22	22,000	+2,000	-2,000
4	20	20,000	22	22,000	+2000	-2,000
5	20	20,000	18	18,000	-2,000	+2,000
6	20	20,000	18	18,000	-2,000	+2,000
7	20	20,000	20	20,000	0	0
8	20	20,000	20	20,000	0	0

Several things need to be noted here. The first is that that B's receipt (or A's payment) is merely p'q, where q is the amount contracted for in the swap – or 1000 barrels in this example. We also see that in this example $\bar{p}$ is the same for both transactors. This need not be so. As shown by an example in the next section, having different $\bar{p}$'s for the counterparties can exaggerate the spread through which the intermediary makes its profit.

In addition, something needs to be said about the market price (or prices) In a sense the actual prices (p_{MA} and p_{MB}) are irrelevant to the actual swap transaction, since cash transfers are based on the index p'. Furthermore, as indicated earlier, the actual amounts being bought and sold by the transactors do not have to be the same as the notional swap amount. In fact, one of the criticisms bought against MG from various persons in academia was that MG had hedged too large a proportion of its throughput.

One final item before turning to a more complicated swap arrangement. In the present study of derivatives, we use two versions of the basis. One of these is the difference between the spot and futures price at a given point in time, while the other refers to the difference between the price (at a given point in time) of similar (but not identical) items – such as WTI oil and Brent oil. A swap employing a WTI index might not be an adequate hedge for UK North Sea production, where the 'marker' price is Brent, although it could turn out to be satisfactory at least some of the time. If this type of arrangement is resorted to, the transactor should be aware of the basis risk.

Although it is not widely appreciated, there can also be basis risk of the second kind mentioned above with identical commodities. For example, in the United States natural gas market, inadequate delivery capacity can result in substantial price differences between different regions. This kind of basis risk makes the US gas futures market considerably less efficient than many observers realize.

At this point the reader is probably ready to admit that swaps are much easier to understand than futures and options. This is particularly true when we have a situation of the type discussed above, where B – who is a buyer of light crude – walks in the front door of a financial institution and wants a swap involving a fixed price of X, for Y years, and involving Z barrels of light crude, while A – who is a seller of light crude – walks in the back door of the same institution, and wants a swap involving the same commodity, price, maturity, and amount. In line with the previous discussion, B might pay fixed and receive floating, while A receives fixed and pays floating.

But what about those situations when buyer and seller have completely different requirements in so far as quantities, maturities, etc are concerned. One way this can be dealt with by a swap arranger is to present the transactors with a 'swap menu'. For example, if the price (index) of crude today is $20/b, the intermediary might offer the following swap quotes:

Swap Maturity	Fixed Swap Price to B($\bar{p}$)
6 months	$20.75
12 months	$21.50
18 months	$22.50
24 months	$24.00

The logic here should be obvious. Assume that B elects a six month swap, and the index price (p') remains at 20. Then, if payments are on a

168

quarterly basis, B pays G $0.75/b at the end of the third and sixth month. Suppose, however, that immediately after entering into the agreement, the price jumps to $22/b. If it stayed at that level, G would have to make quarterly payments of $(22.00 − 20.75)/b = $1.25/b to B. Something that needs to be emphasized here is that swap menus of the above variety are almost certainly negotiable, and thus even under ideal circumstances this market would probably lack the visibility of a futures exchange.

Continuing, it could happen – sooner or later – that G became unhappy with its prospects, and decided to start looking for a counterparty. For example, a player who wants to guard against a price fall. Calling this transactor A, we might find him being offered (to begin with) the following menu:

Swap Maturity	Fixed Swap Price to A
6 months	$19.50
12 months	$19.00
18 months	$18.75
24 months	$18.25

Now let us start the exercise over, with both B and A 'buying' 6 month swap arrangements. Should the price jump to $22/b, and stay there, then G would pay B $1.25/b at the end of each quarter, while receiving $2.5 from A. Thus, in certain circumstances, being a swap arranger can be a very profitable activity. As remarked by an American bank executive: "Any bank which effectively reduces oil price risk will make a tremendous windfall."

Next let us examine, on a non-technical level, the concept of a *participation swap*, in which e.g. a buyer of oil wants to obtain the protection of a *cap* – i.e. a maximum price – but at the same time be in position to enjoy the benefits ensuing from downward price movements. On the other hand, a producer of oil might want to be able to guard against a downward price movement, while benefiting from a rising price. In both cases this strategy would require two comittments:

Producer's Participating Swap: (1) An oil swap plus (2) a 'long cap'

Consumer's Participating Swap: (1) An oil swap plus (2) a 'long floor'

Note what these arrangements involve. For the producer the swap itself will provide a floor, while if the price rises, the long cap – which is nothing more than a call option – results in revenue for the producer to the extent that the oil price exceeds the exercise price specified in the cap agreement. An opposite situation prevails for a consumer, with the swap providing a ceiling (i.e. cap), and the long floor – which is a put option – providing revenue to compensate for the disadvantage of a fixed swap price ($\bar{p}$) in the event of a price decline.

Clearly, there is an explicit hedging cost associated with participating swaps in the form of the purchase price for a cap or floor. Accordingly, the transactor may purchase only a fractional cap or floor. This fraction is usually called the 'participation ratio'. The larger the participation ratio, the greater the explicit hedging cost (i.e. option premiums), although failing to make adequate provision for sharing in large (implicit) profits can often be a bad career move for the lady or gentleman in charge of hedging.

Before going to the conclusions, it might be appropriate to provide a partial explanation of the huge losses experienced by MG on their swap positions.

The explanation turns on the rollover losses resulting from having to renew short dated swaps in a situation where, more often than not, new swap contracts became progressively more expensive. The reason for this, the intermediary might have argued, is that if the term structure of oil prices was examined, it showed a rising trend (i.e. contango), and as a result these circumstances had to be incorporated into the pricing of a swaps deal.

A question must then be posed: why didn't MG lengthen the maturity of their swap transactions? The answer can be deduced from the hypothetical swap menus shown above: on occasion, longer-dated commodity swaps can be very expensive. Moreover, since many observers were of the opinion that, on the average, the term structure of oil prices would shown backwardation, then it was pointless to become involved in an overly complicated hedging strategy. Instead, an ordinary, unsophisticated, 'no-brainer' should produce the desired results. (Backwardation, remember, is a declining term structure, and quite common in various oil and oil products derivatives markets).

In the case of futures contracts, conventional wisdom suggests that, ceteris paribus, it is possible to hedge over long time horizons by rolling over short-term positions, but these futures should be spread over many delivery dates, rather than being 'stacked' (i.e. concentrated) in a particular delivery month, which MG's hedging team unfortunately deemed appropriate. Since, as pointed out above, swaps markets are, to a certain extent, analogous to futures markets without speculators, this logic probably applies to them also. In addition, there was the not-so-trivial matter of MG providing 'cash-out' options with what amounted to zero premiums for buyers of the petroleum products MG was selling.

However, even with these judgemental deficiencies, the opinion here is that bad luck was almost as important in explaining MG's difficulties a bad market diagnoses. Had prices in the markets MG was involved with functioned the way they 'statistically' function, then MG's directors and shareholders might have been spared a great deal of bad news.

6.8 Conclusions

Some readers will undoubtedly want to know what a long chapter on the derivatives markets is doing in a book on energy economics. The answer is as follows.

I know of only a few major energy conferences in the last decade in which derivatives were not considered. At the same time, I have never attended one of these conferences in which derivatives were adequately treated. I also seem to recall the celebrated news program '60 minutes' devoting almost 20 minutes to these instruments, and getting most things wrong – to include blaming the Orange County (California) *contretemps* on derivatives, when actually most of the trouble was due to the amateurish use of 'repurchase agreements' (or 'repos' as they are called) for bonds.

And so, like it or not, a knowledge of derivatives is not just useful, but essential to the education of energy economists.

6.9 APPENDIX

6.9.1 A-1. A Simple Futures Market

It might be useful for readers whose curiosity about derivatives markets has been aroused to examine the following artificial, but pedagogically useful, example of a futures market with 3 traders – speculators X, Y, and Z – who are involved in transactions on 3 trading days, but cannot strike deals the rest of the month. This example should provide an insight into the activity known as marking-to-the-market, or daily price resettlement. But to fully grasp this concept, the reader must comprehend the significance of margin.

For each contract bought or sold, traders must post a small percentage of the nominal price of the contract with their broker. This amount, which is a kind of deposit, is known as margin. (Often this margin can be posted in the form of short term securities, and so the trader does not forfeit any interest income). Contracts are then revalued or marked-to-the-market after the close of each trading day, at the market's closing price. If the closing price has moved below the previous day's closing price, then traders who have bought contracts (i.e. who have long positions) might be asked to pay additional margin; while those who have sold contracts (i.e. have short positions) will show a profit on their contracts that can be taken in cash, without closing out their positions. This process ensures that profits and losses are not carried to far forward.

The numerical example referred to above is presented in the following table. Here I list the trading day, prices (P) at which sales (S) or purchases (B) of futures contracts take place, in dollars/barrel; the volume of transactions (V); and open interest (OI). Open interest, which is extremely important, is defined as the number of open (i.e. not closed) contracts, bought *or* sold, at the end of a given period – but not both. Particular attention should be paid to the entries signifying marking-to-the-market. These are MM_x, MM_y, and MM_z, where a minus sign indicates losses, and a plus sign shows profits. The contract being discussed is for delivery in December, with transactions taking place in November. The last trading day of the month is the 29th, and to keep things simple, initial margin is taken as zero. All trading is for one contract (= 1,000 barrels).

Table A-1: Fictitious transactions for three speculators

Day	P	X	MM_x	Y	MM_y	Z	MM_z	V	OI
1	30	B-long	-	S-short	-	-	-	2	1
10	25	-	(-5)	B-long	(+5)	S-short	-	2	1
29	35	S-short	(+10)	-	-	B-long	(-10)	2	1
			+5		+5		-10		

On day 1, X buys and Y sells a contract. The number of transactions here is taken as two, since in reality both X and Y deal with the clearing house of the exchange, rather than with each other. (Often, however, this is regarded as only one transaction.) On the other hand, open interest is unambiguously unity.

Now let us assume that there is no trading until day 10. At the end of this day, Y closes out his position with a purchase, while Z opens a position. X, by way of contrast, does not buy or sell, but because the price of futures contracts has fallen, the marking-to-the-market of X's constract means that he owes his broker $5/b (= $5,000 on a single contract). Y's contract is also marked to the market, but since he is closing his position, this simply means collecting the profit that he has made on his position.

Finally, let us move to day 29, the last trading day of the month. The price is 35, and X and Z close out their positions by, respectively, selling and buying a contract: the marking to the market shown in the table indicates that X gained $10/b while Z loses $10/b. X's total gain is thus 5 (= −5+10). Notice also how the total losses and gains of X, Y, and Z offset each other: 5 + 5 − 10 = 0. This confirms the zero profit characteristic of a clearing house, where for every 'winner' there is a 'loser'. Almost all futures exchanges have associated clearing houses. This institution acts as seller to buyers, and buyer to sellers – which means that it functions as an intermediary in the transfer of monies between traders (or, to be more exact, the brokers of traders.) Normally, individual traders do not have contact with the clearing house.

In case readers are wondering how can we have a viable market with just three 'players', one fictitious answer might be as follows. The three are sitting in the first class lounge of an airport, and prices are being announced over the PA system. These prices originate on a genuine exchange with thousands of traders. Our three elitists, however, only trade with each other – either directly, or via their brokers, who are easily reached via cellular telephones.

6.9.2 A-2. Rolling a Hedge Forward

As far as I can tell, it is the comparatively low level of theoretical energy economics that has permitted the discussion of Metallgesellschaft's (MG's) 1993 oil hedging strategy to remain a hot topic, and it will probably remain a generous source of academic contemplation and puzzlement until well into the 21st century. The key point here is that when you are running hedge positions for huge amounts of money, interest rates should be brought into the picture in a meaningful way.

Even though (in theory) you break even in the long run when hedging with futures – assuming no basis risk, and ignoring the time value of money – if the margin calls start coming in, and really big dollars are involved, hten rational decision makers must decide whether to pay these calls (with money that could be loaned or otherwise invested), or to shut the position down. As it happened, MG's management, advisors, and very likely creditors decided that it was time to head for the emergency exit.

MG was in the following situation. As traders, they were obtaning and selling oil and oil products, and what they seem tohave been pincipally concerned with was hedging their short positions in commodities with purchases of futures. (of course, almost nobldy bothers to note that two-thirds of MGs hedge position was in OTC swaps.) these futures contracts ewere of short duration because, contrary towhat you may have been told on CNN or when you studied finance, there is insufficient liquidity in many markets to make large scale, long-term hedging an attractive proposition.

We can now introduce some notation. At each 'node' we see the pair $[P_{ii}, P_{ij}]$. With all prices referring to one barrel of oil or oil products, P_{ii} is the price of a long futures contract at time 'i', that was opened at time (i-1), and which has an expiry (or maturity) date of 'i'. Similarly, P_{ij} is the price at time 'i' of a contract that is opened at time 'i', and matures (i.e. expires) at time 'j'. If there is no basis risk, then P_{ii} is equal to the spot price at time 'i', or P_i.(This is due to arbitrage at or just before time 'i': if, for example, $P_i > P_{ii}$, then the purchasers of futres would take delivery and sell on the spot market, which would reduce P_j.) The following diagram looks at the long hedging situation for a few periods, beginning at t=1. As noted, the first transaction involves *buying* a futures, and so the sign is negative (-).

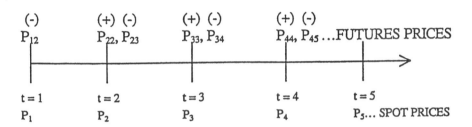

At e.g. t=2, the (long) contract opened at t=1 is closed, and more or less immediately a new position is opened that costs P_{23} (per barrel). This is what 'rolling over' or 'rolling forward' the position is all about. The gain G for that and subsequent periods is

$$G = (P_2 - P_{12}) + (P_3 - P_{23}) + \ldots + (P_T - P_{T-1,T}) \qquad (A1)$$

In this expression T is the (unknown) terminal date for the hedging program. Now, without basis risk, we have e.g. $P_{22} = P_2$ etc, and so we can write for (1):

$$G = (P_{22} - P_{12}) + (P_{33} - P_{23}) + \ldots + (P_{TT} - P_{T-1,T}) \qquad (A2)$$

In examining this expression we see that G is unequivocably greater than zero if in all periods we have $(P_x - P_{xy}) > 0$ or $P_x > P_{xy}$. Moreover, for any period, this inequality signifies a condition that is known as *backwardation*. In formulating its strategy, MGs hedging team simply noticed that on the oil market, backwardation had prevailed 70 – 75% of the time over the previous dcade. As a result, they felt that rolling forward their hedges in the manner described above was certain to result inlarge profits.

Where did things go wrong? Well, according to Professor Ross, the market is too efficient to allow profits to be made so easily. (What he musthave mant though is that profits cannot be made so easily for a firm whose futures exposure was 60—65 million barrels of oil, but with a play of a million dollars or so (every quarter) in the same market, it might have been a different story.) Somewhat less conjectural, the comparatively small percentage of the time when the market was in *contango* (with $P_{xy} > P_x$) instead of backwardation was, for various reasons, transformed into a situation where contange became the rule. *Ceteris paribus,* the effect of this transition on G in equation (A2) should be obvious. Ross also intimates that certain traders figured out that it was possible to take advantage of MG's being in a position where they were literally forced to roll over their expiring positions every few months. This is a very interesting observation, because in a 'textbook' futures market which features adequate liquidity, that kind of insight would undoubtedly have been useful, but it would not have resulted in MG leaving \$1 – 1.5 billion on the table. Interestingly enough, the question that should have been asked has never been asked: why didn't MG forget about futures, and do all their hedging with swaps – assuming that this was possible?

That brings us back to where we began. The simple economics above should have been in general circulation years ago, and might have been if it were better appreciated that economics is an observational science, and down-to-earth theory has more to offer both management and our political masters than pseudo-scientific mumbo-jumbo.

6.9.3 Exercise

1. Look at equation (2) above and assume that contango prevails in every period. Explain in detail what happens to G!

Chapter #7

Electricity and Economics

As most readers of this book are already aware, many topics were broached earlier that apply, in one way or another, to electricity. The reason they were introduced before this chapter is that most students of economics do not want to be subjected to an overdose of jargon and doctrine having to do with such stimulating themes as power-plant loading and capacity factors.

Having been an engineering student, I certainly can respect this attitude; however something that none of us can escape is the overwhelming importance of electricity, or the great liklihood that it is going to become even more important. For instance, a few years ago I was ceremoniously informed that in many circumstances it is a mistake to go through the process of using a fossil fuel to generate electricity, instead of making direct use of the fuel. This is because conversion efficiencies are not particularly impressive: in some countries it appears that for every kilowatt-hour of electricity produced, as much as three or four kilowatt-hours of a primary fuel (e.g. oil, coal, or gas) may be necessary. In these circumstances, it appears that for all its convenience and ubiquity, the cost of electricity is actually very high.

Unfortunately, this is the wrong conclusion, because if we take a careful look at what has happened in countries like Sweden and the United States, we should be able to see the crucial role played in economic development by electricity. I say 'careful' because Swedish economists have focused on the substitution of machines (i.e. capital) for labor, without recognizing that the kind of productivity increases taking place in Sweden featured the substitution of machines *and* (electrical) energy for labor. The way this took place is as follows.

While energy intensity did not steadily increase with respect to output in Sweden (and many other countries), it did grow with respect to the use of capital and labor. It was a lack of familiarity with the details of this phenomenon that led certain observers to believe, mistakenly, that output could be raised indefinitely while the energy intensity of production either stagnated or fell. What happened was that any declines in the energy intensity of output could be accounted for by technical change – largely activated by energy intensive inputs – increasing output by so much that, percentagewise, aggregate output increased by more than energy consumption. (Note: the energy intensity of output = energy/ units of output. A similar expression holds for the energy intensity of input).

Even more important, economists like Sam Schurr of the Electric Power Research Institute have shown that electrification has meant a flexibility in industrial operations that would have been impossible with any other form of energy, and this was the principal reason for the growth in productivity referred to above. In the future very large amounts of electricity will be essential for the optimal employment of computers and robotics; and something that everyone seems to have overlooked is that enormous quantities of power may be required to transform biological and unconventional resources to motor fuels.

7.1 Some Introductory Remarks

It has already been pointed out (in Chapter 2) that the annuity formula is an extremely important tool. Suppose we use it to get some idea of the *capital cost* for a nuclear reactor, where the *investment cost* (I) is 2,000 dollars/kilowatt; the discount rate is 9%; and the amortization period – which in some cases might be regarded as the 'effective length of life of the equipment' – is twenty years. In other words, n = 20. (Note that the amortization period is the time in which the equipment should be paid for. The actual length of life of a well maintained plant is usually considered to be about 40 years). We then get for the capital cost P_c:

$$P_c = \frac{r(1+r)^n}{(1+r)^n - 1} I = \frac{0.09(1+0.09)^{20}}{(1+0.09)^{20} - 1} (2000) = 219 \, dollars \, / \, kilowatt \, / \, year$$

Naturally, this could have been written dollars/year. Let us also notice that the discount rate in the above, which is fairly low, might be considered a 'fair' rate of return on invested capital for public utilities and similar firms subject to price regulation. These firms are more or less assured that they will obtain this rate of return.

We now have the capital cost of a unit of equipment having a *power* rating of one kilowatt. What we want is the capital cost of a unit of *energy*, which is obtained as follows. If the one kilowatt slice of machinery given here functioned 24 hours per day, every day of the year, then we would be talking in terms of 365 (days) x 24 (hours/day) = 8,760 hours. The 219 dollars would thus be spread over 8,760 hours, and so the capital cost would be 219/8760 dollars per kilowatt hour, which is more simply expressed as 2.5 cents/kWh. Remembering that one cent is 10 mills, we get 25 mills/kWh. (Note: mills is a special expression that is not universally used.)

But we cannot expect this machinery to be perfect, and function every hour of every day of the year. Conseqently the *capacity factor* – or actual number of hours the equipment is in service in relation to the total number of hours in a year – must be introduced. (This capacity factor is sometimes confused with the *load factor*, but the load factor is usually defined as the ratio of average daily deliveries to peak-day deliveries over a given time period.) With a capacity factor of 0.75, the number of hours that the above piece of equipment operates per year is reduced to 0.75 x 8760 = 6,570 hours. The capital cost of energy is thus increased to 33.3 mills/kWh.

But observe, the definition of capacity factor used by e.g. the US Energy Information Administration (EIA) takes as the period of time considered a year *minus* the period every year that is used for 'normal' maintenance, to include refueling. Thus, on the basis of EIA practice, the top ten nuclear units in the world have capacity factors that are above 98%. Five of these are in the US, and three in Japan.

The chances are that this 33.5 mills/kWh would not be low if compared to the capital cost for e.g. oil and coal – assuming, of course, that we use the same discount rate and amortization period; but on the other hand the fuel costs for nuclear facilities should be lower than that for these other energy resources. Whether it pays or not to use this equipment will then depend on several things that will not be elaborated on here because they vary from country to country. These things include operations-maintenance costs, 'waste' disposal costs, and environmental costs of various descriptions. But also the size of the capacity factor: in the example above it was 0.75, but if it could reach 0.85-0.90, as it does in 'best practice' installations, that would make nuclear energy very attractive from a cost point of view.

I doubt, however, if even higher capacity factors would enhance the popularity of nuclear energy with its opponents – which is a good thing. The greatest danger posed by this source of energy is complacency on the part of its advocates and operators. In an era when the demand for high profitability is escalating, there is a real prospect that, at some installations, inspection, maintenance, and general alertness will be reduced to a minimum in order to raise shareholder spirits and managerial salaries.

We now come to an elementary presentation of a very important matter: the optimal plant mix. This will be given an algebraic treatment later on.

First of all, there has been a widespread tendency among economists to ignore the optimal plant mix. This is unfortunate, because what we always see in the real world is the utilization of a mix of electricity generating techniques, even though the application of conventional demand curves to so-called electricity supply curves could suggest that – in the long run at least – only one generating technique is required. For example, in countries like Japan and France, it might be possible to argue on the basis of mainstream supply-demand analysis that all electricity should be generated by nuclear facilities, since the marginal cost of this electricity is appreciably lower than that obtained from coal and gas. In general, this argument is wrong.

What we need to understand here is the crucial importance of the configuration of demand: the demand for electricity typically varies during a day in the cyclic pattern shown in Figure 7.1a. This is a much different proposition from the highly stable demand curves that we deal with in our courses in microeconomics, and also vastly different from the scheme that we see in Figure 7.1b, which occasionally appears in important official documents, where it is used to draw conclusions that are presented to politicians and their advisors as scientifically incontrovertible.

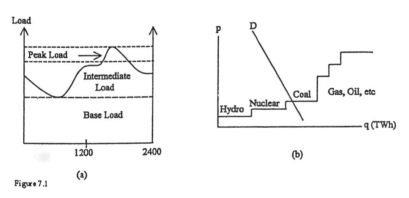

Figure 7.1

Along with the demand contour, it is necessary to understand the different cost characteristics of various types of generating equipment. For instance, even though the marginal running cost of nuclear equipment is very low, while that of gas might be comparatively high, it can be easily shown that it is generally more economical to use gas for generating the electricity required during peak load hours, than to use nuclear capacity. This might also be the appropriate occasion to say more about Figure 7.1b, which shows a conventional demand curve, and a more-or-less conventional supply curve. The usual interpretation of the arrangement in this diagram is that no electricity should be generated with oil or gas, but as already suggested above, and which will be proved below, such an interpretation is highly misleading. More important, this kind of diagram has only a minimal utility where the present subject is concerned.

We shall continue by postulating that nuclear equipment has the highest fixed cost, while natural gas has the highest variable cost. In reality, 'hydro' has the highest fixed cost, and the lowest variable cost; while in some cases combined cycle equipment may have substantially lowered the variable cost of electricity produced with gas via increases in the efficiency of fuel conversion. In any event, until recently the fixed cost-variable cost situation had the following appearance:

FIXED COST	VARIABLE COST
Nuclear	Gas
Coal	Coal
Gas	Nuclear

These cost ladders have the following interpretation. Nuclear and coal facilities are costly to build and equip, and as a result we do not want them standing idle a large part of the time. Consequently, they (and hydro) are prime candidates for supplying the base load, which is the portion of the load that is on the line for 24 hours per day. On the other hand, with comparatively inexpensive gas turbines that can be constructed in small units, and that can be easily switched on and off – but whose fuel costs might be comparatively high – the ideal role is supplying the peak load. As we see in Figure 7.1a, that load is on the line only a few hours every day. To a certain extent the same thing is true for oil, but in electricity generation oil has become the energy source of last resort for most of the industrial world. (Although, even in e.g. Sweden, in those periods when demand is exceptionally high, oil burning electricity capacity will be switched on.)

We can thus understand that while a textbook supply-demand scheme might have the appearance shown in Figure 7.1b, which implies – incorrectly – that gas and oil are 'inferior' inputs, real world generating systems involve, on the average, about 60% baseload generation, 30% intermediate, and 10% peaking. (This particular average should be regarded as what is called in the United States a 'ballpark estimate', and for some countries may not be applicable). At the same time it should be appreciated that as more combined cycle equipment is introduced, and its efficiency increases, it will play a larger part in carrying the base load. At the Futsu plant of the Tokyo Power Company (Tepco), 14 gas turbines are combined with 14 steam turbines to produce 2,000 MW of power at a rated efficiency of about 41 percent, and apparently a large part of this is base-load power.

Something else that needs to be emphasized is that because of the impossibility of predicting demand, all real world electricity generating systems must contain a substantial reserve capacity, which often takes the form of older oil based equipment, with a high operating cost. Remember that where electricity is concerned, the slightest overload – i.e. demand exceeding supply – is not followed by an increase in price, but by a blackout or brownout.

7.1.1 Exercises

1. The last sentence in this section may pose a problem for some of your colleagues. Why don't you frame an explanation for them!
2. Later it will be indicated that gas fueled electrical generating equipment has a lower capital cost than nuclear facilities. Why wasn't that reflected in the 'supply' curve in Figure 7.1b?
3. What is the difference between a discount rate and an interest rate?
4. In the annuity calculation given earlier in this section, what happens to P_c as 'r' decreases? What happens to it as 'n' decreases? Can you provide 'plain language' explanations for these results?
5. In 1995, per-capita electricity consumption in the US averaged 12,000 kWh/y, while in the world as a whole this was 1,200 kWh/y, and in the developing countries 800 kWh/y. What does this amount to in barrels of oil per person in a perfect generating system? Roughly, in an actual generating system?

7.2 Daily Load Curves and Load Duration Curves

Most of the electricity generated in the industrial world originates in non-nuclear steam, hydro, and nuclear power stations. In the United States non-nuclear steam predominates, while in Sweden it is nuclear and hydro. Australia has no nuclear power plants, and mostly relies on non-nuclear steam, with coal being the principal input for raising steam. Italy still uses considerable oil to produce steam. Etc.

As a medium for doing work, steam appears to have been introduced about the year 200 BC. The first practical steam engine was developed by Thomas Newcomen in 1705, but it was James Watt who improved this invention to a point where it could truly be called one of the technological wonders of all time. For the efficient driving of electrical generators, however, a radical modification of the steam engine was necessary. This turned out to be the high-speed turbine, which is superior to the steam engine for driving electrical generators.

The power of flowing water may have been harnessed 5,000 or 6,000 years ago, but it was during the period of the Roman Empire that water power was extensively developed in order to free, for military use, horses that were being used for such things as driving corn mills. The device that originated from this situation was the waterwheel which, in its traditional form, was only of limited efficiency. A few millenia later water turbines were developed which could be used to drive generators in hydroelectric plants, and do so at a very high efficiency. Water power used to generate electricity is not subject to the kind of losses which occur in steam engines, where for thermodynamic reasons the theoretical maximum efficiency is 60 percent – although in practice the best thermal efficiency that has been recorded in commercial installations not employing a 'combined cycle' system is about 40 percent.

When available, hydroelectric plants are generally used to generate the base load, but because the output of these facilities is so easily raised and lowered, they could also be useful for supplying some peak load requirements. Furthermore, a combination of steam and hydro power plants favors the use of flexible operating schedules which can greatly improve system performance, since when the stream flow is high the hydro component can increase its output, while during periods of low runoff, steam facilities can generate a larger share of the load.

The first large commercial nuclear plant went into operation in the UK at Calder Hall, Cumberland, in October 1956. What we generally have with this type of installation is a nuclear reactor producing steam that drives a turbine, that in turn generates electricity by driving a generator. Nuclear power plants are exclusively intended for generating the base load and, perhaps, a portion of the intermediate load. Their capital costs are too high for them to be idle more than a small fraction of the time, while at the same time their operating costs tend to be relatively low.

A few remarks are also in order on the gas turbine power plant. This type of equipment basically consists of a compressor, which draws in air and raises its temperature; a combustion chamber where fuel is added to the compressed air and burned; and a turbine which is driven by the hot combustion gases. Finally, the turbine drives a generator. Because of its rapid start-up capability, the gas turbine is regarded as ideal for supplying peak load power.

Now we can turn to the important matter of power plant loading, though on a non-technical level. As shown in Figure 7-2a below, the power demand on a generating station can vary widely during a 24 hour period. Naturally, this demand also depends on the time of year, as well as the structure of the load: industrial, commercial, residential, etc. In countries like the US the power load can be extremely heavy during in the summer because of the pervasive use of air conditioning; while in Sweden, where air conditioning is not widely used, industrial operations are greatly reduced during the summer months, and lighting and heating requirements are relatively low during the same season, the smallest loads of the year are usually registered sometime between the first week in June and the first week in September. The depiction in Figure 7-2b shows a typical electric power load duration curve that is derived from Figure 7-2a. Please note, however, that Figure 7-2b is an approximation, in that it assumes the *daily* curve in Figure 7-2a remains unchanged for the entire year, which is highly unlikely. (The two small figures belong to Exercise 1 at the end of this section).

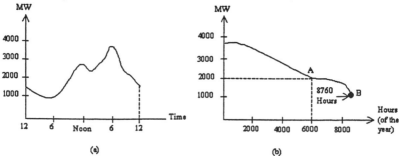

(a)   (b)

Figure 7.2

The scheme in Figure 7.2a should be self-explanatory, but it is likely that the power plant load duration curve in Figure 7.2b requires some discussion. The way I handle this is to construct a very simple numerical example below in order to demonstrate how this curve is obtained, although even here it is an easy matter.

The diagram in Figure 7.2b shows that we always have a load of at least 1,000 MW on the line. Thus, roughly speaking, the base load is 1,000 MW. At the same time, the capacity of the system must be sufficient to satisfy the largest load that might appear, which is about 3,900 MW. In addition, there should be some reserve generating capacity in the event of unplanned outages due, for example, to equipment failures, and which could be coupled in during periods of routine maintenance.

Instead of the annual load duration curve shown in Figure 7.2b, a daily load duration curve could be drawn (and for pedagogical purposes this is done later), but annual load duration curves seem to be the rule. In interpreting Figure 7.2b, observe again that at least 1,000 MW is on the line every hour of every day in the year, while at least 2,000 MW is on the line 6,000 hours. It was mentioned that the base load was 1,000 MW which, usually, would be generated by equipment designated base load equipment; but as will be shown, even that part of the load which is not in the base load – for example, the part represented by the stretch B-A in the diagram – might be generated by such equipment, depending on the size of this load, and the relative cost of equipment deemed suitable for this mission. In general, this means all equipment except that specifically dimensioned to accommodate the peak load.

In the context of the discussion so far, it is possible to consider several well known parameters. The first is the load factor, which will be defined here as the ratio of the average load on a power plant during a certain period of time, to the peak load occuring during that period. (Personally, I am not completely satisfied with the peak load being designated here. In certain circumstances it might be useful to use the average load over e.g. XYZ in Figure 7.1a.) In Figure 7.2a the total load on the power plant over a 24 hour period is the area under the power load curve, and so actually it is total energy demand. Thus, the average load is this value divided by 24. Written out, we get for these relationships:

Average Load = Total Load/24 (= Total Energy Demand/24)   (7.1)

Load Factor = Average Load/Peak Load   (7.2)

A less useful, and conceptually less satisfactory parameter is the *demand factor*, which will be mentioned here only for completeness. It can be written as:

Demand Factor = Maximum Load/Total Connected Load   (7.3)

Occasionally, so-called daily or annual plant factors are computed. These are defined as the ratio of the electrical energy produced during the period of interest, to the product of the number of hours per period (e.g. 8,760 hours if the period is a year) and the total rated generating capacity of the entire installation.

$$Daily\ Plant\ Factor = \frac{Electrical\quad energy\quad produced\quad per\quad day}{(Hours\ Per\ Day)x(Rated\ Generating\ Capacity)}\quad (7.4)$$

Now let us examine a simple numerical example designed to illustrate some of these concepts. A power plant has an installed generating capacity of 1,000 MW, and a connected load of 900 MW. On a certain day the peak load is 600 MW, and the electrical energy produced during that day (i.e. the area under the electric power load curve) is 7,200 MWh. We thus see that the Average Load is 7200/24 = 300 MW, while the load factor is 300/600 = 0.5. This load factor is not particularly high. A high load factor is desirable

184

because it indicates an economical utilization of capacity. Put another way, very high peak loads over short periods of time are undesirable. As for the demand factor, assuming – as is often done – that maximum load = peak load, we have 600/900 = 0.67 for the demand factor.

The demand factor is normally in the range 0.4-0.6. Finally, the plant factor is 7200/(24x1000) = 0.3. A high daily plant factor is desirable because it indicates the economical use of generating equipment.

Earlier, we introduced a parameter called the capacity factor, which seems to be especially useful for nuclear equipment. A more comprehensive definition than that given earlier defines the capacity factor as the ratio of the electricity produced for the period of time considered, to the energy that could have been produced at continuous full-power operation during the same period. The thing here is to be clear about the 'period of time considered'. This was taken as a year minus the normal 'down time' for maintenance and refueling. Thus it turns out that *load factor* and *capacity factor* are not at all the same thing.

Finland is a country where high load and capacity factors are commonplace; but if we think in terms of the amount of nuclear based electricity produced in relation to total electricity production, then Sweden has a very impressive nuclear sector. It should also be emphasized that the US has a number of high quality installations, with capacity and/or load factors that match or exceed the best in the world. If the worst American installations were eliminated, and some of the remainder given the kind of upgrading that was once suggested for the reactors of Eastern Europe, then it would be very difficult to prove that nuclear power in that country was more expensive than power generated in coal or gas based installations. It has been suggested, however, that from time-to-time 'workmanship' in the US nuclear sector has left a great deal to be desired.

We can now take a closer look at the construction of the load duration curve, and here, for pedagogical reasons, I am proposing a new kind of day – a day with only 6 hours. The first hour is 0-1, and the last hour is 5-0. The face of a clock for this strange day is shown just below a table giving the loads (in kilowatts, and *not* kilowatt hours) that apply to the given intervals. Power load and load duration curves have also been drawn in Figure 7.3.

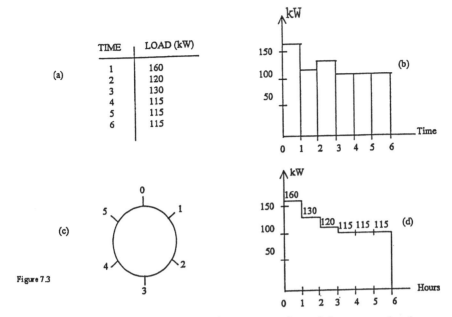

TIME	LOAD (kW)
1	160
2	120
3	130
4	115
5	115
6	115

Figure 7.3

A comment might be useful on the construction of the power load curve. Readings of power consumed were perhaps taken 'on' the half hour (e.g. at 1:30), and the value of the load on the line at that time (= 120 kW) registered as the load on the line for the entire period (i.e. period 2). Clearly, the load may vary over this period, but if a more accurate description of the power being demanded is desired, then more frequent readings would probably be necessary.

Once we have the power load curve, it is a simple matter to construct the load duration curve shown in Figure 7.3d. As indicated previously, this construction usually covers one year, but here the operation was, uncharacteristically, performed for a period of one day. The reader should have no problem in going from Figure 7.3b to Figure 7.3d, although perhaps he or she should observe that a load equal to or exceeding 120 kW is on the line 3 hours a day.

If you have some problem with the above, think about it in terms of your house or apartment. Suppose that it contains 10 lamps, and no other electrical items. At mid-morning you might find one or two of them on, occasionally; while in the evening, when you are studying from this book, you might have seven or eight on. But, note, we have only talked about power, in terms of the wattage of these lamps. To get energy a time dimension is necessary.

Accordingly, our next step is to calculate the area under one of the curves in Figure 7.3. For instance, if we take the power load curves we get:

$$E = \Sigma t_i(kW)_i = 1 \times 160 + 1 \times 130 + 1 \times 120 + 3(1 \times 115) = 755 \text{ kWh} \quad (i = 1,...,6)$$

7.2.1 Exercises

1. In the above expression the index (i) runs from 1 to 6. Indicate this arrangement using the summation (Σ) sign instead of (i = 1,...,6)!
2. In Figure 7.3 the figures are for one day. Using these daily figures repeat this exercise for a period of one year. Discuss!
3. Assume that the daily load curve, every day of the year, is as shown in Figure 7-3b. Sketch a load duration curve for the year!
4. Suppose that the total connected load in Figure 7.3 is 200 kW. Calculate the average load, load factor, and demand factor! In general, can you distinguish between load factors and capacity factors?.

7.3 The Economics of Load Division

We can now turn to the crucial matter of showing how the load is divided between several types of equipment, employing the concepts introduced in previous sections.

The first thing that we need is an annual load duration curve. A typical load duration curve is shown in the bottom frame of Figure 7.4a, and in Figure 7.4b I have shown a replica of one of these curves for Norway in 1975. In looking at these figures the reader should attempt to recall the algebraic expositions in previous chapters. In the upper figure of 7.4a we have three linear equations: $F_1 + tV_1$, $F_2 + tV_2$ and $F_3 + tV_3$. As we remember from one of these expositions – or from secondary school algebra – an expression such as $K = F + tV$ represents a linear equation, with F the intercept on the vertical axis and in this case (with 't' the variable on the horizontal axis) V is the slope (or granient) of the expression. Consider also the possibility of obtaining the (K,t) coordinates of the intersection of e.g. $K_1 = F_1 + tV_1$ and $K_2 = F_2 + tV_2$. From these two equations we get $t = (K - F_1)/V_1 = (K - F_2)/V_2$ and from the last two expressions *you* can solve for K. Once you have K, you can put it in either expression to get t. (Exercise! Using $F_1 = 5$, $F_2 = 15$, $V_1 = 1$ and $V_2 = 0.5$, solve for the values of K and t when these two equations intersect, and draw a rough sketch which shows the value of F for each of these expressions, notes the point of intersection, and approximates the Vs. Finally, add another $F + tV$ expression, and repeat the calculation. Please note that the figures given are not very realistic.)

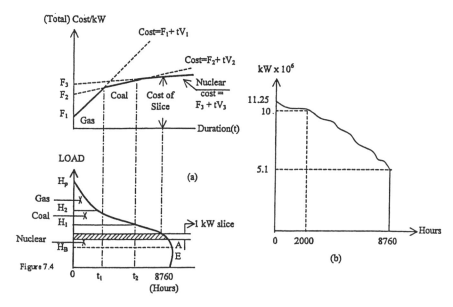

Figure 7.4

Before moving to some rather simple algebra, certain things need to be made clear about this diagram. We have three different types of equipment: nuclear, coal, and gas. It would be very convenient indeed if we could say that nuclear generates (or 'carries') the base load, coal the intermediate load, and gas the peak load; but if we stick to the concept that the base load is the load that is on the line all the time, and only this load, then nuclear equipment probably generates more than the base load. Perhaps, then, it is best to put it the following way where this and similar diagrams are concerned. We have base load plant (nuclear in this example), intermediate load plant (coal), and peak load plant (gas).

There is also nothing inevitable about the configuration. There are systems in which combined cycle gas equipment generates the base load, and conceivably others in which it generates the entire load – base, intermediate, and peak. It is also usual in Sweden (and elsewhere) that there are installations containing several independent nuclear 'units'. In terms of our present discussion, one of these may have a capacity factor approaching 100%, while another unit may have a palpably lower capacity factor since its output will fall under rated (full power) capacity when demand declines.

In Australia, plants that are designated intermediate load plants comprise about 40-45 percent of total installed capacity, as compared to 50 percent for base load plants, and 5-10 percent for peaking plants. Generally, the peak load represents a very small segment of the demand for electricity, and usually only a fraction of this capacity is in use, but since the available generating equipment must be able to satisfy the maximum demand that may

188

appear in the system, capacity or load factors for peak load facilities are very often under 5 percent. Consequently, as much peak load power as possible should be purchased rather than generated in the system, and as a result many utilities do everything possible to buy their peak load requirements from firms in other cities, states, or countries. But even so they must have reserve capacity, since it could happen that there is a simultaneous peak in surrounding regions.

The upper frame of Figure 7.4a is meant to show the total cost (capital *plus* variable) of employing 1 kW of capacity over the period t. Obviously, if t is small, then it makes sense to use equipment with a low capital cost, even though fuel costs may be high. Gas turbines are traditionally the technology of choice for this role, as indicated in Figure 7.4a. For instance, to the left of t_1, gas is preferable to coal and nuclear. But after t_1, the high cost of the gas itself outweighs the low capital cost, F_1, of gas turbines, and from a cost point of view, coal becomes a more satisfactory fuel. Similarly, for loads that are on the line longer than t_2, nuclear is optimal. As mentioned earlier, and shown in the upper frame of Figure 7.4a, nuclear technology has a high capital cost, but a comparatively low fuel cost. Now, with F as capital cost, and V as variable cost, and specifying that $F_3 > F_2 > F_1$, and $V_1 > V_2 > V_3$, it is surprisingly easy to calculate the cross-over points.

To begin, recognize that the total cost of supplying a load of 1 kilowatt for t hours of the year using a plant of type 'i' is $F_i + t_i V_i$, where the units are:

$$(Dollars/kW) + hours \times (Dollars/kWh) = Dollars/Kilowatt$$

As is clear from Figure 7.4b, regardless of the number of kilowatts that are on the line, if we have constant unit costs then gas turbines are the most cost efficient equipment for supplying loads having a very short duration. But eventually the time comes when $F_1 + t_1 V_1 = F_2 + t_1 V_2$. This can be immediately solved to obtain $t_1 = (F_2 - F_1)/(V_1 - V_2)$. It is also the case in Figure 7.4 that at some point we are going to obtain $F_2 + t_2 V_2 = F_3 + t_2 V_3$. As in the previous case, this expression can be solved to obtain $t_2 = (F_3 - F_2)/(V_2 - V_3)$. Continuing, in order to see how these expressions function, let us examine some figures provided by the Tucson Gas and Electric Company in 1975:

Type of Plant (Equipment)	F	V
Base	100 (F_3)	0.0024 (V_3)
Intermediate	40 (F_2)	0.0200 (V_2)
Peak	20 (F_1)	0.0275 (V_1)

It is therefore possible to calculate that $t_1 = (40 - 20)/(0.0275 - 0.0200) = 2,666.6$ hours, while $t_2 = (100 - 40)/(0.0200 - 0.0024) = 3,409.0$ hours.

We also need to show that, for this exercise not to be trivial in concept, the cost curves in the upper frame of Figure 7a intersect to the left of $t = 8,760$. As it happens, this will be the case if we have e.g. $F_3 - F_2 > 8760(V_2 - V_3)$, which can be easily proved. Electrical engineers sometimes call these curves *screening curves*.

This might be a good place to note that in one strongly nuclear oriented country, France, the cross-over point when substituting nuclear for coal (t_2) is probably less than 3,000 hours. The reader should also remember that there is a difference between the structure of an electricity generating system in capacity (kW) terms, and in terms of the actual amount of energy generated. For instance, in 1990 about 50 percent of the French system was nuclear in kW terms, but nuclear energy provided almost 75 percent of that country's electrical energy (in kWh). Another interesting characteristic of the French system is the large difference between marginal costs in base load periods, when most of the load is being generated by nuclear equipment; and periods in which peak units with high running costs – such as oil based units – must be utilised. It has been claimed that the (short run) marginal costs for these two extreme situations can differ by a ratio of 20 to 1.

The French electricity bureaucracy is unreservedly partial to LMC pricing, but in most other countries, power company directors have fallen madly in love with SMC pricing. This has been alluded to earlier, but it might be a good idea to say a little more about this topic.

If we examine the type of diagram that many of us habitually use to discuss this topic, Figure 7.5a, then it appears that SMC pricing does not have much to recommend it. Ignoring for the time being the inconvenience for analytical purposes of multiple technologies, we see that we have exactly the sort of thing that shareholders (and bond purchasers) would never tolerate: with price (p) = v, there would be intolerable losses, although, perhaps, these could be decreased by the use of large entry charges. LMC pricing could be a better deal, although again entry charges are usually necessary.

190

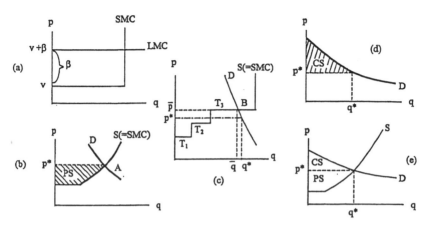

Figure 7.5

But what happens if we have the kind of SMC curve that you worked with in your first economics course, Figure 7.5b, and which is probably more suitable for this analysis. This is also a supply curve (if we limit ourselves to the upward sloping part of an SMC curve that lies above the average variable cost curve), and if it is intersected at e.g. 'A' by a demand curve, substantial profits might be realised. The conclusion that should be drawn at this point is that if real life supply curves take on the appearance of the one in Figure 7.5b, or 7.5c, then the construction in Figure 7.7a is a very unhappy approximation.

Looking at Figure 7.5b again, the cross-hatched area is defined as the producer's surplus (PS), which is closely related to the profit: we can write Profit = PS − Fixed Cost, and so in most situations PS can be used as a kind of proxy for the profit. The curve in Figure 7.5b could also serve as an acceptable approximation of the supply curve for a multiple technology firm, because it does not provide us with any debatable specifics about the techniques being employed to produce q, although for the purpose of the present exposition, Figure 7.5c is more useful.

If we have average cost pricing, which would be the usual departure under traditional regulatory practices, the price at which $\bar{q}$ – or to be more exact, q^* – units would be sold might be p^*: under regulation, only normal profits are allowed, and thus, algebraically, it appears that some of the output in the diagram would have to be sold at a loss in order to balance the (economic) profits made as a result of the low cost production as represented by technologies T_1 and T_2. (But notice the term 'algebraically'! In the real world, deliberately selling a good at a loss would not normally be a part of standard operating procedure unless strategic considerations were involved.)

However, if we had SMC pricing, $\bar{q}$ would be sold for $\bar{p}$, and that includes most of the output produced with low cost capacity. ('Most', because some output would normally be sold under long term contracts for less than $\bar{p}$, and perhaps much less.)

We can now ask if the power companies (outside France) have anything against LMC pricing? The apparent answer is no. In the UK, for instance, the equilibrium price of electricity is supposed to converge to the long run marginal cost for the purpose of encouraging the supply of new generating facilities; however these days everybody's favorite subject is privatisation, and according to theoreticians like the late Fred Schweppe of Massachusetts Institute of Technology, SMC pricing is an intrinsic part of the privatisation package.

As we have seen in the UK, however, the important thing is not what the price is, but what it does. It is supposed to keep shareholders happy, consumers reasonably pleased, and the salaries and share-options of directors climbing – a phenomenon that some observers have termed 'the new socialism'. The key element in spreading so much joy does not, up to now at least, appear to be what happens to prices, but to costs. Costs are adjusted by getting rid of as many employees as possible, while convincing those who are retained to increase the volume and effectiveness of their work, and if they have any attitude problems to keep them well out of sight, or to have them cured at the first opportunity.

A few more comments on the above subjects should round out this part of the presentation. Producer's surplus involves the difference between the price of an item, and the variable cost of producing it; while consumer's surplus has to do with the difference between the 'value' of an item (as measured in Figure 7.4d by the distance between the horizontal axis and the demand curve) and the price paid for that item. This is p* in the same diagram (but not the same p* as in 7.4d). For instance, on a hot day you might be willing to pay 3 dollars for an ice cream cone, but if you only had to pay 1 dollar, your consumer's surplus was 2 dollars: the difference between value and price.

Usually consumer's and producer's surplus are shown in one diagram, such as Figure 7.5e. As you have found out, or will find out in your advanced courses, the neo-classical optimum condition p = SMC emerges from the mathematical process of maximizing (CS+PS), where this sum is sometimes called 'total welfare'.

Will this price cover costs? If not, then an entry (or usage) cost would logically be called for, and usually entry costs are resorted to regardless of the profit outlook. In the textbook situation that we are discussing here, one strategy is to charge an entry fee that extracts enough consumer surplus (measured in e.g. dollars) to nullify any losses that result from setting p =

SMC *or* p = LMC. This is easy in the classroom, but in practice it might require an unattainable precision.

Finally, we can say a few things about electricity pricing in the UK, because some observers believe that practices in the UK should serve as a model for the rest of the world. As mentioned, if the demand for new plant is to be satisfied, then the LMC should be brought into the picture in a meaningful way. Unfortunately, very few persons seem to know how this should be done, and much academic work is concerned with refining the p = SMC rule, mostly along lines suggested by Crew and Kleindorfer (1979) in their well known book. In its simplest form, the p = SMC rule (for 'wholesale' electricity) is transformed to:

$$P = SMC(1 - P_R) + V_L P_R \qquad (7.5)$$

P_R is the loss of load probability: the probability of capacity being insufficient to meet demand. V_L is the value of the lost load: the customers' value of the opportunity cost of outages, or benefits foregone through interruptions in the supply of electricity, and these benefits are likely to be much larger than the price of a marginal unit of electricity. (Part of the logic here is that V_L should be such as to provide a reward for making capacity available in the short term, even if this capacity is only seldom required; while in the long term it should influence the decision to construct new capacity, or to upgrade older and perhaps more inefficient plant.)

Since in practice P_R will almost never be zero, then according to (7.5), P = SMC becomes a curiosum where the sale of 'wholesale electricity' is concerned. The actual 'price capping' formula in the UK, as it is called, is pondered over by the regulator – or 'reregulator' – and his advisors, and their conclusions imposed on electricity distributors. It is not (directly) determined by the market, although presumably its selection is influenced by market portents of one type or another. The relationship is:

$$\Delta P/P = RPI - X \qquad (7.6)$$

This merely says that electricity prices are allowed to increase at a rate ($\Delta P/P$) that is equal to the percentage increase in the retail price index (RPI) minus a factor (X) which, in the mind of the regulator, represents improvements in 'efficiency' of one kind or another.

It is rather difficult to make a scientific evaluation of (7.6), except to say that it is not the economics equivalent of Einstein's theory of general relativity. It can be reported, however, that since its introduction electricity shares in the UK – despite a few setbacks – have shown a tendency to outperform the *Financial Times* general index, and while the nominal (i.e.

the money) price of electricity has increased, the real price may have decreased somewhat. Productivity has, of course, gone up on all fronts in the newly deregulated/reregulated sectors, but with many tens of thousands of persons being 'retrenched', and thousands of others expecting a dose of the same, it was only to be expected that output/head would increase.

7.3.1 Exercises

1. As stated in the text, the condition for an intersection of cost curves to the left of $t = 8,760$ is $F_3 - F_2 > 8760(V_2 - V_3)$. Can you show exactly what it means if we have $F_2 - F_3 = 8760(V_3 - V_2)$ and $F_1 - F_2 = 8760(V_2 - V_1)$?
2. (Extra credit) Suppose tht we have 3 technologies (T_1, T_2, T_3) as in Figure 7.5c, and that fixed cost was zero. This means that profit is equal to the producer's surplus. Why? Now, suppose that the demand curve is perfectly inelastic, and intersects the supply curve at B. Put some numbers in this figure, and calculate the profit using SMC pricing! Also calculate the total revenue and total cost! With the same numbers, calculate – at least approximately – what P* would have to be if we had average cost pricing!

7.4 Some Final Observations on Electricity Tariffs

For residential consumers in the US, rates often depend only on the amount of energy consumed per time period (e.g. month or quarter). However, a declining block pattern is sometimes followed, with successive blocks of energy having a lower charge per kWh. The intention here is to recover some fixed charges (such as those for metering) in the first, or high tariff, block. By way of contrast, for major commercial and industrial consumers, there can be a separate charge designated to specifically cover the expense associated with rather complicated and expensive metering. For example, the monthly tariff of the Fall River Electric Light Company, in 1980, called for a 70 dollar charge for the first 25 kW of capacity, or less, a 1.70 dollar/kw charge for the next 275 kW, and so on. Energy charges showed a similar configuration: the first 30,000 kWh cost 1.04 cents/kWh, the next 50,000 kWh cost 0.64 cents/kWh, etc.

Before continuing this discussion, the reader should examine Figure 7.6, where an outline of a typical electricity supply system is presented.

194

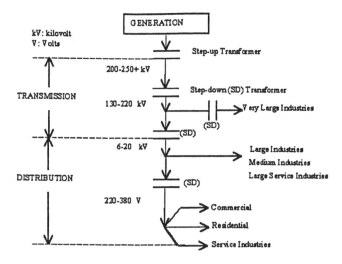

Figure 7.6

It tends to be the case that industrial consumers of electricity in many countries are charged less for electricity than other consumers. This is because industrial consumers generally have high load factors, and for the most part consume electricity at higher voltages than commercial or residential consumers, which can mean considerable cost savings on such things as step-down transformers and the equipment associated with low voltage distribution lines. Another point of considerable relevance for this discussion is that when price discrimination can be practiced by a seller of electricity, those consumers having inelastic demands are typically charged more than those having elastic demands. Those having especially elastic demands would be, typically, firms with the possibility of generating their own electricity. For instance, the US aluminum industry probably generates more than 30 percent of the electricity it uses.

It should also be emphasized that in a time of considerable deregulation, large industrial consumers appear to be greatly favored as compared to smaller consumers, since they can negotiate the electricity price with several suppliers, and choose what they consider to be the most favorable. But at the same time this remarkable advantage can turn sour. The following brief tale deserves to be carefully scrutinized. It helps to explain why economics is a fascinating, but sometimes puzzling subject.

Industrial consumers in the UK, and elsewhere, are among the major supporters of deregulation. After all, they still have the right to negotiate with electricity suppliers, but in addition they have access to the power pool, which functions as a (day ahead) spot market, and where – they have been told – deregulation will ensure that prices are lower than they normally would be on the fixed price contracts that they customarily negotiate with generators. Sometimes this turned out to be true, but sometimes things turned out quite differently. In the middle of 1999, for about 12 hours a day, power pool prices were approximately five times as large as during the previous month, and since these prices are used as a benchmark for fixed price contracts, no immediate respite was available in the 'fixed price market'. Instead, large power users such as steel and chemical companies found their profit margins squeezed, and their competitiveness undermined. The story here was that deregulation unambiguously increased the market power of generators, and when the opportunity presented itself they took advantage of the situation: according to the Energy Intensive Users Group, the generators "manipulated" prices. The pool sets half-hourly wholesale prices in England and Wales, and in some of these half hour periods price spikes were so severe that the most energy intensive firms could not afford to buy power, and simply interrupted production.

Mainstream economic theory does not seem to approve of different electricity tariffs for residential and industrial consumers, which goes to show how wrong mainstream economic theory can be some of the time. In Sweden, the price of electricity is clearly perceived by all political parties to be a macroeconomic variable, and it is generally believed that the large export industries should be insulated from any unpalatable changes in the level of this price. Instead, they will almost certainly devolve on households and small business. What we have here is an important example in theoretical welfare economics: whenever significant changes take place, there will be winners and losers, and in situations where losers cannot be adequately compensated, it is politics rather than economics which dictates the initiation and extent of these changes.

An expression that one sees quite frequently these days is *wheeling*. From an engineering point of view, wheeling is simply the transmission of power, and mostly 'bulk' power; but administratively it means a situation where the establishment doing the transmitting is neither a buyer or seller of this power, but is selling the transmission service. Thus a typical wheeling operation involves three transactors.

Perhaps the first comprehensive tariff designed to cope with the problem of the variation in electricity demand was the so-called 'Green Tariff' (Tarif Vert) of Electricité de France (EdF), which received its name from the color of the cover of the brochure in which the tarif was described. A decade after its introduction, this tariff became compulsory for all high-voltage electricity consumers, and at the same time available for residential consumers in a modified version. Conceptually, the background for the tariff was as follows.

Since the Paris area is the largest consumer of electricity in the country, although at some distance from most *generating* sources, the Paris price was the basis of the pricing system. The intention was that at any hour of the day the short run marginal cost of electricity delivered to the Paris area would be taken as being equal to the *variable cost* for an extra unit of output from the least efficient plant supplying Paris with its electricity. The expression 'least efficient' is very important here, because if the output of electricity is allocated among plants in such a way as to minimise cost, then the marginal kilowatt-hour of electricity will be taken from the least efficient plant of those in operation.

Elsewhere in France costs might be lower due to the availability of hydro-electric power, but since Paris is capable of absorbing all the hydroelectric power available, and would still need power generated from some other source, the cost of hydroelectric power could not be taken as the marginal cost even in the region where it was generated, because the marginal cost in those vicinities should, from the theoretical point of view, be related to the opportunity cost, which is the price that would be paid for it elsewhere. Instead it was stipulated that consuming areas closer to hydro facilities than Paris should pay a price for electrical energy equal to the marginal cost at Paris *minus* the transmission cost between them and Paris.

This kind of thing might make it appear that Paris is being favored over the rest of the country, but the opposite would be closer to the truth. EdF is one of the most popular firms in France, especially among its employees, with a proved record for creating employment, and using revenues from, e.g. Paris, to make sure that electricity is available to consumers in out-of-the-way localities, and at tariffs virtually equal to those paid by citizens living on or near the 'Boul Mich'. Notice also that the SMC was mentioned above. This is not, however, the price of electricity. That price is roughly the LMC (= SMC + some capacity charges) plus, very likely, some 'financial charges'.

This does not, however, make French electricity more expensive than that sold in neighboring countries. On the contrary; and it might be even more inexpensive in the short run if the directors of EdF listened to the directors of the European Commission, and initiated a program of deregulation that featured the firing of a few tens of thousands of employees, at least. Notice the expression 'short run'! In the long run French electricity would be sold in the regions bordering France, where the price is generally higher, and when that begins to happen nobody really knows how high the price of electricity to French households would go

We can bring this chapter to a close with a few remarks about rate of return regulation. If we forget all about taxes, depreciation, and that sort of thing, then the required revenue requirements R of a regulated firm can be written R = r*K + V. In this expression r* is the allowed rate of return for the firm; K is capacity (in monetary units) – i.e. the value of the capital used to produce the output q, and V is variable cost. Now we can go the the price (p) that the producer is allowed to charge for q, which is:

$$P = (r*K + V)/q \tag{7.7}$$

All this should be self-evident, except that where do we get a chance to mention the *rate-base*, and the much talked about (in the technical literature) 'Averch-Johnson effect'?

The rate base above is K: essentially the value of the generating company's plant and other assets that are *actually* serving its customers. Note the emphasis on 'actually', because occasionally one hears sensible arguments for putting at least some expensive unfinished (i.e. non-producing) capacity into the rate base, particularly if there is a shortage of capacity. Presumably, this would serve as an incentive for new investment.

As for the Averch-Johnson effect, this suggests that if the allowed rate of return (r*) on capital exceeds the regulated firm's true cost of capital, then it is very profitable for the firm to expand capital by more than the 'optimal' amount – i.e. the amount characterised by the isoquant-isocost tangency that you encountered in your basic course. This is immediately implied in the equation just above, where it is hard to think of a situation in which it would not be in the interest of utilities managers to present regulators with the largest possible K.

The opinion here is that the Averch-Johnson effect reduces to a detail, at best.

7.5 Conclusions

Among the many topics that could have been taken up in this textbook, but which were omitted or treated only perfunctorily, hydroelectric power probably deserves special mention. Here we have the damming of flowing water, which is then directed through turbines of high efficiency and low speed, that are connected to generators. Electricity generated in these installations is low cost electricity, and a partial proof of this is that Norway, Eastern Canada, and Sweden have considerable hydro-electric capacity, and also the lowest cost electricity in the world. As it happens, environmentalists are generally opposed to new dams and the inundations they require.

In the case of virtually all electricity generation technology, the basic mechanism displays a turbine prime mover driving a rotating electric generator. This applies to 'hydro', and it also applies to the other end of the size scale. There we have the gas turbine-compressor combination originally developed for aircraft, and which eventually may operate at very high efficiencies. The problem is, however, that no matter how efficient the machine, it cannot function without primary energy resources (to include water in the case of hydroelectricity). Neither cleverness nor the sweat of one's brow can substitute for a supply of 'potential energy' (in the form of oil, gas, etc) that can be transformed into mechanical work or its equivalent.

Much has also been said earlier about the last observation, although perhaps one thing should be added. I once attended a lecture in which one of the most brilliant economists of the 20th (or any other) century, Nicholas Georgescu-Roegen, flatly stated that most major wars of the last 150 years were about natural resources, to include energy resources. Professor Georgescu-Roegen may not have been completely correct, but his warning is something that should never be forgotten.

Chapter #8

Uranium, Nuclear Energy, and an Introduction to Intertemporal Production Theory

Much of this chapter concerns uranium, but nuclear activities are not ignored. Electricity production from nuclear facilities was 7% higher in 1998 than 1997, and in the US the Department of Energy's research budget has just been raised by 13 million dollars. The US accounts for 25% of the world's CO_2 production, and it appears that the US Congress has concluded that without the continued presence of nuclear energy, the US cannot fulfill the Kyoto agreements. The nuclear industry has responded by suggesting that the 'life' of some installations should be extended from 40 to 60 years.

As is well known, nuclear energy is not a particularly popular medium. Even in France, virtually the capital of 'the peaceful atom', there is no shortage of persons who hope that someday another energy source will take over. Frankly, unless there is a major accident in that country, or a nearby country, this yearning seems unrealistic. In countries like France and Japan, where energy independence is paramount, nuclear energy is not there to be questioned, but to be used.

It also needs to be made clear that the next accelerated wave of nuclear construction will likely commence a very long time before the New Year's Eve parties begin on January 31, 2199. I was recently informed by a referee of one of my papers that my concern about population growth was completely unjustified, and perhaps he or she was correct; however unless a particularly nasty version of Malthusian logic comes to pass, I very strongly doubt whether a world population of less than 8 billion souls can be avoided by the year 2050, and if this turns out to be the case, then nuclear plants could spring up faster than discos were being opened in New York and Paris during the swinging sixties. (World population has now reached 6 billion,

200

which means that a doubling has taken place in slightly under 40 years. In these circumstances, 10 billion by 2050 is probably a better estimate.)

At the present time, as much as one-sixth of the world's population might lack domestic access to electricity in any form, although some observers put the figure much higher. In any event, percentagewise, this fraction is probably increasing. As a result, when representatives of the anti-nuclear movements inform various governments in the Third World that their energy requirements can be met with sun, wind, and wave, they are politely shown the door. Without adequate (i.e. large) supplies of energy, genuine economic development and progress is will be extremely difficult to achieve.

On the other hand, renewable energy sources have a crucial role to play in the energy picture of the 21st century, and the world is going to be in a bad way if they do not become widely available in a few decades. But one strategy for making sure that they will not be available is to continue to insist that they are already economical, if only our political masters would come to their senses. The ladies and gentlemen who sit in our elected assemblies may often have their minds on other matters than the welfare of their constituents, but they would have to be comatose not to notice that in countries like Sweden, where every conceivable effort has been made to develop alternative energy capabilities, the result has not been promising. This might also be a good point to say something about the cost of nuclear energy. Swedish nuclear energy is taxed at almost 4 times the average elsewhere – the tax comes to 39.1 (Swedish) öre per kWh, plus an extra tariff of 2.2 öre that is not registered as tax; but even so the price of electricity in Sweden is about on the same level as Norway, where more than 90 percent of the electricity is generated using water power, which is probably the most inexpensive energy medium.

Just as Sweden wants to turn away from nuclear energy, some countries are welcoming it with open arms. For instance, as indicated in Table 8.1, Asia is on its way to establishing an inventory of nuclear plants as large as that of Europe, and may reach this level by 2010.

Table 8.1

(Year) COUNTRY	1992 Capacity	1992 Number	2000 Capacity	2000 Number	2005 Capacity	2005 Number
Japan	41,600	51	45,500	54	47,100	56
South Korea	8,600	10	13,700	16	18,700	21
Taiwan	5,100	6	5,100	6	7,700	8
China	2,100	3	2,700	4	4,300	8
India	2,000	10	3,300	15	3,800	16
Pakistan	140	1	460	2	460	2
		81		97		111

Source.
OECD/IEA

In case anyone is interested in the real nuclear dilemma, it is that although development work on 'safe' nuclear equipment is slowing, thanks to 'market uncertainties', the total number of nuclear reactors continues to expand, and eventually could escalate everywhere in the world. As a result we could find ourselves with a huge stock of (from the safety point of view) sub-optimal equipment, and possibly a renewed interest in the breeder reactor. When that happens, a full-fledged plutonium community is right around the corner, which is something that that the world will probably not be ready for soon, and may never be ready for.

8.1 Some Nuclear Background

The first commercial nuclear reactor was constructed in the UK in 1956. It was a gas cooled reactor, and initially attracted considerable attention in France. Eventually, however, the French decided to adopt pressurised water equipment. That country began planning its nuclear sector immediately after the second world war, and openly stated their intention to construct an electricity sector that was based on what many French scientists considered the most important scientific achievement of all time – the nuclear reactor. (Professor Dennis Gabor was also of that opinion.)

Most nuclear facilities are privately owned, although the French nuclear sector, and the very large Canadian firm Ontario Hydro, are government owned and operated – as are nuclear facilities in Eastern Europe and North Korea. About half of the nuclear capacity of Sweden, Finland, and the UK is government owned, while private firms dominate in the remainder of the world.

The most common reactor is the light water reactor (LWR), of which the pressurised water reactor (PWR) is one variant. The other is the boiling water reactor (BWR). The actual reactor use can be considered an interior part of the nuclear fuel cycle, not just because the spent fuel can be reprocessed and, in certain circumstances, significant amount of fissile materials can be recovered and eventually recycled, but because the storage and management of wastes is such an important activity.

In general, reactors are designed to use uranium fuel which is about 2.7-3.3 percent U-235, but U-235 constitutes only about 0.71 percent of natural uranium (or about 1 part in 140). Thus the 'front end' of the nuclear cycle is concerned with upgrading (or enriching) natural uranium to the required quality (i.e. richness). The 'non-upgraded' portion of U-235 is called depleted uranium, or tails. A.D. Owen (1985) provides a valuable example in which 5.5 kilograms of natural uranium (with 0.711% U-235) becomes 1 kilogram of enriched uranium plus 4.5 kilograms of depleted uranium (i.e. tails) containing 0.2 percent of U-235. The front end consists of the first four stages of the nuclear fuel cycle, and stretches from the mine to the reactor core. The last two stages, whose purpose is to resolve the spent fuel problem, are called the 'back end'.

Reprocessing is one of two approaches to spent fuel management, and it involves chemically separating the plutonium, uranium, and radioactive residues found in irradiated fuel. The alternative is storing spent fuel in untreated or partially untreated form. There is a genuine economic dilemma here due to the difference in cost of the two options, particularly since these costs are dependent on forecasts of future fuel supply, and the efficiency of waste management.

Energy produced from fossil fuel is the result of an uncomplicated chemical process. On the other hand, energy produced from nuclear fuel originates in the force binding the constituent parts of the fuel's atoms together, and its release necessitates the alteration of the structure of the atom itself. This can be an extremely complex process.

The nucleus of an atom contains protons and neutrons, and the larger this nucleus, the easier it is for these elements to 'disperse'. Uranium is so important because one of the heaviest (and most complicated) atoms in nature is the isotope 235 of uranium (U-235). (Different isotopes of an element occupy the same position in the periodic table, but they do not have the same weight.) This atom contains 235 'elements', and if it absorbs one additional neutron, it becomes unstable, and can divide into two or more fragments (i.e. atomic nuclei), in addition to several neutrons. The mass of these fragments and neutrons is now somewhat less than that of the original nucleus and, most important, the reduction in mass has been transformed into kinetic energy (i.e. motion), which in turn is converted into heat as the fission products collide with surrounding atoms and are brought to rest. Other U-235 atoms may absorb the neutrons released by a previous fission, and themselves undergo fission. A release of neutrons leading to further fissions constitutes a 'chain reaction'.

What we have now is a mass-to-energy conversion in which the amount of energy that can eventually be released is huge. In theory, the energy output of 1 kilogram of U-235 is equal to that contained in 2,000 metric-tons (or tonnes) of oil, which is 2 million times as much per unit weight. Only 0.71% of natural uranium is U-235. The remainder is U-238, which is unsuitable for the direct production of energy, because it captures neutrons without undergoing fission. This would greatly reduce the total energy produced by a given amount of natural uranium (U) were it not for another phenomenon. After capturing a neutron, the nuclei of U-238 transmutes into an unstable chemical element that is not found in nature: plutonium 239 (= Pu-239). Plutonium can also undergo fission, and thus both U-235 and Pu-239 are fissile elements, while U-238 is called a fertile element. The process by which a fertile element is converted to a fissile element is called breeding, and it will be mentioned later

The uranium ore that is mined usually contains less than 1% uranium. This ore is milled to produce a concentrate that is between 85% and 92% uranium oxide (U_3O_8) that is called yellow cake. This is followed by further purification and conversion to uranium hexaflouride (UF_6), which is usually designated natural uranium. (Note how we have distinguished between uranium ore and natural uranium).

In natural uranium the ratio of fissile to fertile atoms is 1/140, which implies some difficulties in maintaining the fission process. Slowing down neutrons with the help of a non-absorbent medium called a moderator will increase the probability of a neutron being absorbed by a U-235 nucleus, although the most applicable method for enhancing fission is simply to use enriched uranium. This involves boosting the U-235 content of UF_6 from 0.71% to at least 3%. The enriched product is then converted into uranium dioxide (UO_2) pellets, which are further processed into a form (e.g. fuel rods) which can be inserted into a reactor core. Now, with fission induced heat available, this heat can be utilized to produce steam, which is eventually transformed to the motive power for a conventional electricity generator. The power output is controlled by varying the population of neutrons in the reactor core. (It might also be useful to know that heavy water reactors (HWR) of the Canadian CANDU (Canadian deuterium uranium) type differs from the equipment mentioned earlier, since its basic fuel is natural uranium instead of enriched uranium).

At least nine-tenths of all uranium used at present is enriched, and the enrichment stage accounts for about 50% of the cost of the nuclear fuel cycle. This activity is measured in separative work units (SWUs), which are a function of the effort required to separate U-235 from U-238. The proportion of U-239 remaining in the depleted uranium (the tails) that results from enrichment is called the tails assay.

A.D. Owen and others have laid particular stress on the importance of the enrichment stage. New enrichment technologies (e.g. centrifuge and laser) have greatly reduced the energy needed to perform enrichment. As a result these technologies display a tails assay of around 0.10% instead of the usual 0.30% (on average). Since about ten percent of the OECD's uranium requirements are met from the supply of enriched tails, the large-scale adoption of these more efficient technologies could reduce the need for uranium by a significant amount. Adopting new technologies is a cost question, however, and these investments are unlikely to be made before the price of uranium begins to escalate.

The back end of the cycle begins with the used (or 'spent') fuel elements being removed from the reactor: a reactor needs refueling when not enough U-235 remains to sustain a chain reaction. After being cooled, they are shipped to a reprocessing facility, where the fuel elements are reduced in size, dissolved in nitric acid, and various fission products are separated out. These are processed in various ways until shipment to a permanent repository becomes feasible. The thing to be aware of here is that the fuel has used up only a small amount of the total energy it contains. There is still some 'unburnt' U-235 in the residue, and from an energy point of view, the plutonium in spent fuel contains at least 50 times the amount released by the 'burning' of the original load.

A 1,000 MW installation can annually produce 15,000 cubic meters (= 15,000 m^3) of low-level waste, 1,500 m^3 of intermediate-level waste, and 20 m^3 of high-level waste. The high-level waste is difficult to neutralize. The first step is often vitrification (i.e. enclosing the waste in glass, which is then enclosed in steel containers). Final disposal might consist of burying these containers in deep holes in geologically inactive areas, etc. Eventually it may be possible to solve the nuclear waste problem by developing a reactor that completely burns up all its fuel.

Uranium is mined in either underground or open pit installations from ores having a concentration that averages about 0.25 percent, if the concentration is measured in terms of uranium. It is sometimes measured in terms of uranium oxide, but this could cause some confusion because uranium oxide can be considered a pocessed form of uranium,

The above numbers imply that about 400 tonnes of ore must be removed in order to obtain 1 tonne of natural uranium. The general abundance of uranium is about 4 parts per million (= 4ppm) in the earth's crust, and 3 parts per billion in seawater. There are a number of estimates in circulation as to the amount of uranium that can be extracted from both onshore deposits and seawater, but all the figures that I have seen concerning possible extraction strike me as being mostly guesswork, and so they will not be quoted here.

There is probably no point in using valuable space to explore the mechanics of the fast breeder reactor (FBR), except to say that it converts a large part of U-238 – the most abundant isotope, constituting 99.3% of natural uranium – into fissile material, and it does the same thing for thorium. Plutonium is a normal output of most reactors, but in the FBR, under the influence of fast neutrons, U-238 is transformed to U-239, which is a fissionable nuclear fuel. These reactors can also use depleted uranium produced in the enrichment process. This is why, in Japan and perhaps several other countries, the FBR is considered the ultimate solution to the electricity supply problem, although this is not the kind of thing to be sung at the top of one's voice in the local karaoke club. Some of the directors of the Japanese energy bureaucracy have made it abundantly clear that, as far as they are concerned, the conventional nuclear power station, which exploits just about one percent of the energy in its fuel, will ultimately be as extinct as the dinosaur, and must be replaced by the fast breeder, which extracts about eighty percent. They will also tell you, in private, that a Japan that can reduce its energy costs could have a substantial commercial advantage over those countries utilizing high-cost energy.

The price of uranium yellowcake was only eight dollars per pound in 1992, and it began to move up in the middle of the 1990s; but as it turned out, this was a false dawn. By 1998, when the McArthur River project in Canada was in the process of going on stream, uranium prices had slumped to near their lowest level in real terms. This project ostensibly contained the highest grade ore-body ever discovered, and thereby helped to reinforce the impression that there was no chance of the world running out of uranium. Of course, as with most markets, observers concerned themselves with the short – or very short – run. The forecasts dealing with an explosive growth in electricity requirements over the next 50 years were mostly ignored.

Uranium reserves are judged sufficient to support existing and planned production to at least 2010. According to the Japanese Institute for Energy Economics, there is a global reserve base of about 2 million tonnes (Mt) of uranium, but the figure published by the International Atomic Energy Agency is 2.5 Mt. This latter organization also points out that at the present time 582 uranium deposits have been identified, however there have been a number of mine closures since the beginning of the 1990s, and in addition many mines entered the 21st century operating well under full capacity.

A factor that complicates the uranium supply-demand situation is the use of blended-down highly enriched uranium (HEU) from nuclear weapons as fuel for nuclear reactors. In the first decade of the 21st century present expectations are that at least 15% or more of OECD uranium requirements will be satisfied by Russian HEU, and perhaps some HEU originating in the US. At the beginning of the new century, about half of the uranium used in reactors originated in inventories of different kinds – primarily military stocks of one kind or another, recycled uranium and plutonium, enriched tails, and government and commercial stockpiles. These particular sources of supply are far from exhausted, and will be called on for a long time in the future. Processing efficiencies could also increase considerably. As a result, the conclusion cannot be drawn that uranium prices will escalate in the near future. But in the long run, with essentially the present technology, the uranium supply situation may not be as rosy as some observers believe.

Canada is probably, on the average, the lowest cost uranium producing country, followed by South Africa, the United States, and France – or, to be exact, French owned deposits in Africa, and perhaps elsewhere. Australia, however, is a country with enormous potential, and in the last half of 1996, with a precipitous lifting of government restrictions taking place, plans were announced to double output if there was sufficient demand. The largest single uranium deposit in the world is at Olympic Dam, in South Australia; and the following table shows an estimate by the French Institute for Energy of the supply and demand for uranium between 1995 and 2010.

Table 8.2

	1995	2000	2005	2010
Reactor Requirements (tU)	61,700	68,700	71,000	80,000
Production capacity (market economies) (tU)	39,200	37,300	40,400	38,600
Additional Req. (tU)	22,500	31,400	30,600	41,400

Source: AIEA (Paris)

As alluded to earlier, technical developments are capable of greatly changing the fuel supply and demand picture, and so the above figures should be regarded as indicative rather than definitive. Advanced fuel management techniques, higher burn-up levels, and better enrichment equipment should reduce the amount of natural uranium required to obtain a given output.

Contracts have also been introduced in which a given quantity of uranium is sold at prices related to the market/spot price at time of delivery – as is often done with copper and oil. Sometimes these contracts contain a price that could be termed an escalated base price, with the price paid by the buyer being the higher of the two. The disadvantage of this arrangement, from the point of view of the buyer, is that even if the world (spot) price fell, the escalated base price could continue to rise, since it was related to such things as production costs.

Although it is not widely discussed, reprocessed Pu-239 can substitute for U-235 in that it can be mixed with U-238 (in the form of uranium oxide) to provide a fissile fuel known as MOX (for mixed oxides). This activity consumes more plutonium than it creates, and thus is not related to 'breeding'; but even so, the presence of plutonium on the input side of the process greatly offends environmentalists and the anti-nuclear movement. They claim that it is safer and more economical to store spent nuclear fuel than to reprocess it for reuse. The 'mixing' refered to above takes place in a reprocessing plant of the type that the UK government has approved of for British Nuclear Fuels' (BNF) Sellafield complex in north-west England. A great deal of attention has been paid to this installation, and the opinion here is that in the name of safety, the more attention the better.

In Germany, environmentalists have almost succeeded in forcing a Social Democratic government to decommission some nuclear capacity, and in Sweden they may have succeeded: it appears that all that remains is setting the final date. One way that this might happen in these two countries is that the oldest reactors are deactivated, while the remainder continue to run for another decade or two. In these circumstances, the remaining reactors would generate a secure earnings stream from facilities that were amortized years ago. The present intention in Sweden is to replace the deactivated reactors with 'green' power, mostly in the form of wind power. Personally, I have high hopes for wind power, but it is a concept whose time will not fully arrive until the output of these facilities can be economically stored.

8.2 Some Intertemporal Cost and Production Theory

The discussion below follows and extends the presentation of Alchian and Allen (1964). The three chapters in that book in which similar materials are found contain some of the most valuable reading in all elementary economics. In case the reader runs into trouble, however, it might be a good idea to review Chapter II of this book.

The example that will be used here turns on the acquiring of a bulldozer to use in mining uranium ore, although it could also be used to mine coal, or oil shale, or whatever. Our intention is to buy the bulldozer and use it, on a contract basis, to 'produce' 25,000 units a year, for which we will be paid a unit price related to the spot price on the market at the time of delivery. Payment will be made at the end of the year, and costs will also be reckoned at that time. After 2 years we will sell the bulldozer on the 'second hand' market.

208

In thinking about this project, we realize that we have some crucial unknowns. The most important of these is the price we will be paid for the uranium ore. Obviously, it would have made our estimate of the profit (or loss) of this operation a great deal simpler if we could have signed a contract with an ore buyer that specified a delivery price. We also do not know the price that we will get for the bulldozer when we sell it in two years. As a result, when we calculate profit, we are actually calculating the expected profit, or the ex-ante profit, and this may turn out to be very different from the realized or ex-post profit.

We also need to remember some advice given us by an old acquaintence, Professor Bill Lather of the Stockholm College of Economic Knowledge. Professor Lather says that when dealing with intertemporal matters, never forget that cost should be defined as a reduction in wealth, viewed as a present value.

For instance, if we buy a machine for $1,000, and sell it for $1,000 a year later, the (ceteris paribus) change in wealth is not $1,000 − $1,000 = 0 (unless the rate of interest in zero, which signifies the absence of 'time preference'). If the rate of interest (r) is e.g. 10%, then the change in wealth is $1,000 − (1,000/1+0.1) = \$90.9$, and this is a decline in wealth: it is negative, and it is a cost. But if the machine has been acquired for $1,000, and sold for $1,100, then there is no change in wealth − i.e. no cost − because as emphasized in earlier chapters, there is no (objective) difference between $1,000 today and $1,100 a year from now in a perfect market if the rate of interest is 10%. But remember, we buy durable goods − e.g. lathes and TV sets − to produce things, or to consume their services, and although there may be a cost associated with acquiring and using these items, our total wealth (measured in money or in 'utility') might increase.

Our first cost concept, the acquisition cost, will only be treated en passant. If we buy a bulldozer for $5,000, and decide to sell it right away, we may only obtain $4,800. In this situation we can speak of an acquisition cost of $200. The assumption made below is that the acquisition cost is zero: if you buy a bulldozer this morning, but decide to get rid of it before the 'midnight-caller' goes to work, it does not cost you anything.

After that there is a cost, which to begin with might be labled a non-use depreciation cost. Not much will be made of this either, since it can easily be subsumed under depreciation in general, but a short example might be beneficial. Suppose that the non-use depreciation cost is $1.50/day. Thus, the cost of acquiring the bulldozer and keeping it for one week, without using it, is $10.5, since we assumed that the acquisition cost was zero. This means that ceteris paribus that we could sell the bulldozer for $5,000 − 10.5 = \$4,989.5$ after one week.

What I call the ex-ante investment cost of the unused bulldozer is then the purchase price minus the discounted expected value of the resale price. We have said that the equipment loses $1.5/day in value if it is not used, which comes to $1,095 in 2 years, and so the (expected) resale value at that time is $5,000 - 1,095 = $3,905$. Discounted to the present this is $3,905/(1.1)^2 = \$3,227$, and so the ex-ante investment cost of the bulldozer is $5,000 - 3,227 = \$1,773$. The annual rent (or annual capital cost) is obtained by annuitizing this amount over 2 years, and with r = 10%, this amount is $1,021/year. But remember, for the time being the asumption is that the bulldozer is not used!

Now for a very important digression. Suppose the bulldozer did not deteriorate at all, and its resale value was $5,000. In that case the ex-post investment cost is $5,000 - 5,000/(1.1)^2 = \$868$. Turning this into an annuity over 2 years, at 10%, yields $500/y. If you remember what we did in previous chapters, where we did not concern ourselves with depreciation, then this was defined as the annual rental cost of capital. It was simply rK, which in this case is $0.10 \times 5,000 = \$500$/year.

We can now ask what the situation is if you own this bulldozer, and do not have to borrow money to buy or rent it? The answer is that if we ignore the tax laws, then nothing changes. This is because we are dealing with an opportunity cost. If you have $5,000, and r = 10%, you could earn $500/year in interest income. Somewhat more abstract, if you own the bulldozer and can sell it for $5,000, then by keeping it you are giving up an income of $500/year. Accordingly, if you think that this asset will not perform as well as a bank account, and you get no other satisfaction from owning it, then you should give serious thought to selling it.

The next move is to consider depreciation due to use. Unlike the non-use depreciation that we have already discussed, and which we could think of as a kind of obsolescence cost, this could be presented as an operating cost. For instance, suppose you buy a new Volvo, and park it in front of your home for 2 years in order for the neighbors to admire your taste, but you do not drive it. If you want to sell it after those 2 years, most potential buyers would take its age into consideration, even though everything else was precisely the same as it was the day you bought it. Thus, you might normally expect to be offered less than you paid for it, although you realise that a pleasant surprise might come your way due to unusual market conditions – such as a very large increase in the price of new cars.

But if you drove this car from Uppsala to Stockholm every day (about 70 kilometers), and in addition every June you drove from Uppsala up to Riksgränsen (in Northern Sweden) in order to ski and party under the famous 'Midnight Sun', the resale value of the vehicle would obviously decrease.

To say that depreciation in use was an operating cost means that we would have to relate it explicitly to output. I prefer avoiding this complication in this elementary presentation, and instead merely say that during its 2 years in our possession, the bulldozer is expected to depreciate in value by $2,500. (This is both use and non-use depreciation.) Note what this does to the expected investment cost, and the (annual) rental rate! The investment cost becomes, with r = 10%, $5000 - (2,500/(1.1)^2) = \$2,933$. If we annuitize this (at 10% over 2 years) we get for the annual rental rate $1,690/year. Once again let us note that this is an expected value, because in reality we only expect that the resale value of the equipment is $2,500.

Before continuing, let us be absolutely clear about terminology. The purchase price of the bulldozer is $5000, but generally this is labled the investment cost: we are investing in a bulldozer. However if this bulldozer has a resale value, then it could be argued that the investment cost is less than $5000. Accordingly, I prefer calling the purchase price adjusted by the resale value the ex-ante investment cost, because we do not know what the resale value of the bulldozer is until it is actually sold, at which time we could, if we desire, calculate the ex-post investment cost. The ex-post investment cost is not particularly fruitful for the present exercise, but by the same token we realise that catastropic mistakes could be made in estimating future costs and revenues.

That brings us to operating costs. There will only be two of these in this exercise: POL (or petrol, oil, and lubricants), and the wage of the bulldozer operator, who also does maintenance on the equipment. Costs and revenues are shown in the following tableau, along with various calculated values. The key calculations involve annuities, and if necessary the reader should return to Chapter 2, and refresh his or her memory on this subject. The revenues given in the tableau are the revenues from producing 25,000 units/year. Notice that 'costs' have a negative sign.

ITEM	TIME		
	t_0	t_1	t_2
1. Revenue (Output = 25,000 units)		+7500	+7500
2. Purchase price (Bulldozer)	-5000		
3. Resale value (Bulldozer)			+2500
4. POL		-500	-500
5. Wages		-4500	-4500
6. Net receipts	-5000	+2500	+5000
7. Present value (net receipts)	-5000	+2275	+4130
8. Profit (calculated from 7)	+1405		
9. Investment cost (from 2 and 3)	-2933		

10. Annual rental rate (from 9)		-1690	-1690
11. Discounted cost/period (from 4, 5, 9)	-2933	-4545	-4132
12. Total discounted cost (from 11)	-11610		

At this point the reader should go over this tableau, and make sure that he or she understands every entry. Then observe the following. In periods 1 and 2 we have running costs of $5,000, and let us add the capital rental of $1,690 to this. Our net receipts/period are then 810 (= 7,500 − 5,000 − 1,690). If these are discounted back to the initial period, we again get for our profit $1,405.

Something that might puzzle the reader in the last calculation is why the resale value of the bulldozer did not enter the calculation. The answer is that it did: it was used to calculate the (ex-ante) investment cost (item 9), from which the rental rates of capital were determined.

That brings us to the main object of this exercise, which is to bring volume (= total planned output) into the cost picture as well as annual output. What we have in our example is an intention to produce 50,000 units in 2 years, at 25,000 units/year, and this is not at all the same thing from the profit point of view as producing 50,000 units in 3 years, at 16,666 units/year. As it turns out, costs are lower with the latter arrangement, but so are profits.

Before proceeding, some conjecture is in order. We can consider what kind of equipment would be appropriate if we were thinking in terms of a total production of 500 units, or maybe 500,000, still assuming that the rate at which production was to take place was 25,000 units/y. The first of these would take 500/25,000 = 0.02 years (assuming a relevant linearity in the production process), which is the same as 0.02 x 365 = 7.3 days. The second would take 500,000/25,000 = 20 years. In our earlier discussion it was implied that to produce a total of 50,000 units would require a certain type of bulldozer. Where 500,000 units are concerned, an optimal solution might call for a larger bulldozer, or 2 smaller bulldozers used for 20 years each, or 2 bulldozers used sequentially, with each being used for 10 years, etc. Conceptually, it is not a particularly difficult problem, although considerable calculation might be necessary before the optimal equipment selection was made.

More difficult is the situation with 500 units. To produce these we might rent a bulldozer on an hourly or a daily basis, or some combination of these. We might do the same thing with a total output (i.e. volume) of 5,000, although it could turn out that buying a small bulldozer was the best move. In any event, thee are many potential outputs and volumes, and many pieces of equipment, and as a result we might in theory find it useful to construct a diagram of the type shown in Figure 8.1b, where the locus, or envelope, is

212

constructed from a large number of diagrams of the type shown in Figure
8.1a.

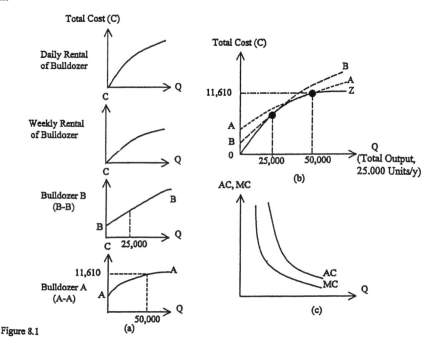

Figure 8.1

Now let us look more closely at the above diagram, and in particular
Figure 8.1b. Q is the total output, or what we occasionally call volume; and
for each value of Q, and a given output per period (or rate of production)
there is an optimal type (and/or quantity) of equipment. For $Q = 25,000$
units/year, bulldozer B might be optimal, while for $Q = 50,000$ units/year,
bulldozer A appears to be optimal. At the same time notice that as we move
to the right along the envelope OZ, the total cost, C, rises, while the average
and marginal costs decrease. This 'flattening out' of OZ implies that
increasing the volume by a factor of two does not double the cost.

Alchian and Allen call this the "economy" of mass, or volume,
production. What they did not show, however, was that part of this declining
marginal cost is a purely mathematical phenomenon. Suppose that we have a
cost X' at the end of 2 years. The present value, C'_0 is then $X'/(1+r)^2$. Now
suppose that we have $X'' = X'/2$ at the end of 1 year. The present value of
this amount can be written $C''_0 = X''/(1+r) = (X'/2)/(1+r) = X'/2(1+r)$. Now
let us compare these:

$$\frac{C_0''}{C_0'} = \frac{X'}{2(1+r)} \Big/ \frac{X'}{(1+r)^2} = \frac{1}{2}(1+r)$$

(or)

$$C_0'' = \frac{1}{2}(1+r)C_0' = \frac{C_0'}{2} + rC_0'$$

In other words, although X' was reduced by one-half, the present value of the cost was reduced by less than one-half. Or we could go in the other direction: instead of X" at the end of one year, we have 2X" at the end of 2 years. The present value of the cost does not double, however.

The next cost effect has to do with annual output, q. In the example we are working with here, 25,000 units/year are being produced for 2 years, at a discounted cost of $11,610. If we increase the rate at which a given volume is to be produced, it is clear that total cost must increase, and Alchian and Allen claim the same thing for average and marginal costs. Their explanation turns on having to bring in more resources in a given period, and thus having to resort to less efficient resources and/ factor combinations. With a few assumptions, some calculus can show this; but that demonstration will not be provided in this book. Instead, the reader can examine Figure 8.2, which shows the total cost situation.

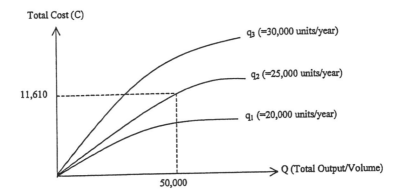

Figure 8.2

The two concepts of cost presented above can be examined more closely by considering proportional increases in volume and output/year. The volume effect, ceteris paribus, would tend to reduce unit costs, while the

214

output effect, by itself, would tend to raise them. By combining these two effects in a single diagram, as shown in Figure 8.3, we get cost as a function of both output and volume.

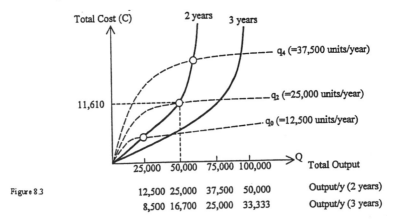

Figure 8.3

		Output/y (2 years)
12,500 25,000 37,500 50,000	
8,500 16,700 25,000 33,333	Output/y (3 years)

Having constructed and discussed, in some detail, a 2 year program, it might be useful to see what a 3 year program looks like. This will be done for a situation where total output remains at 50,000 units, and it is assumed that at the end of 3 years, the bulldozer can still be sold for $2,500, but the operating costs are distributed proportionally over the 3 years. The argument for keeping the resale price of the bulldozer at $2,500 is that the total wear-and-tear on the bulldozer remains the same, since the total output is unchanged, and there may be more time for maintenance.

Obviously, this particular assumption means, ceteris paribus, lower costs and higher profits. As we will find out in the following tableau, however, this effect was not sufficient to raise profits although, had the discount rate been lower, such might have been the case. Once again the reader is invited to examine these figures very carefully, and if possible duplicate the calculations behind the tableau, but employing a much lower discount rate. The discount rate here is 10%.

ITEM	TIME			
	t_0	t_1	t_2	t_3
1. Revenue		+5000	+5000	+5000
2. Purchase price (Bulldozer)	-5000			
3. Resale value (Bulldozer)				+2500
4. POL		-300	-300	-300
5. Wages		-3000	-3000	-3000
6. Net receipts	-5000	+1700	+1700	+4200
7. Present value (net receipts)	-5000	+1545	+1404	+3155
8. Profit (calculated from 7)	+1104			
9. Investment cost (from 2 and 3)	-3121			
10. Annual rental rate (from 9)		-1254	-1254	-1254
11. Discounted cost/period (from 4, 5, 9)	-3121	-3000	-2727	-2480
12. Total discounted cost (from 11)	-11327			

Observe this result. As mentioned earlier, costs went down but also profits. Thus, if we put a cost curve for a 3 year program in Figure 8.3, it may be below the 2 year program for all values of Q, but this does not mean that profits are higher.

Several other things need to be understood here. If we insist on a 2 year program, then an optimal program is one whose volume maximizes total (discounted) revenue minus total (discounted) cost. Thus, the 2 year program we discussed above may not be the optimal 2 year program. Similarly, if the length of the program is 'free', then to optimize profit we should choose both the length of the program and the annual production. Clearly, we are now dealing with something a bit more complex than the isocost-isoquant tangencies of your elementary course.

How well do producer of uranium and other energy resources face up to challenges of this nature. According to Dominique Finon (1989), the evidence is mixed, but he does not see spot prices playing the role they should according to mainstream textbooks. I concur in this, given the role of uncertainty in real life: it seems quite natural to me that contracts must have a larger role in energy markets, and spot prices a lesser role, than indicated in those textbooks.

Nothing would please me more than to end this section right now, but regrettably another topic must be taken up. As noted earlier, we are often dealing with expected values, and perhaps the only way this can be avoided is with ironclad and comprehensive contractual arrangements. Otherwise we are thrown back on our ability to estimate, or guess, what will happen in the future. For instance, if our wage costs at the end of this year are $4,500, then we might assume that they will be $4500(1+0.025) = \$4,612$ at the end of next year, assuming the expected inflation rate to be 2.5%.

But how do we approach this matter of future revenues, given the uncertainty or risk associated with future prices? (Formally, the difference between risk and uncertainty is that with the former, some probabilities can be estimated concerning the likelihood of future events, while with the latter probabilities cannot be estimated, and often it is impossible to define them. For example, the probability of finding substantial substantial quantities of oil in the Chaco, as compared to the probability of finding intelligent life on the planets Mars and Pluto. What we have chosen to do in this book is not to make a great deal of the difference between these two terms, which is the usual practice.)

One way to handle this problem is to arbitrarily add a risk premium to the risk free discount rate (e.g. the government bond rate). In the above tableaux we assumed the discount rate to be 10%, and did not go into the mechanics of selecting this value. However if there was a great deal of money involved, and the particular enterprise was of the high risk variety, then we might decide that a discount rate of 12%, or even more, was better. The word "arbitrarily" might seem out of place in a serious textbook, but when he was at Harvard, the present US Secretary of the Treasury, Lawrence Summers, once found that firms were using 'hurdle rates' of 8-30% in their evaluation of investments, although the nominal interest rate during the period in question was only 4%, and the real rate was close to zero.

Obviously, it would be nice if we could systematise this matter of choosing risk premiums, and one suggestion is to employ the Capital Asset Pricing Model (CAPM). What this involves is comparing the return on e.g. a single investment with the expected return on an investment comprising the entire share/stock market.

This latter return means that you invest in a bundle of shares that replicates the entire market – or, realistically, something close to the market. One possibility would be a mutual fund composed of the Standard and Poor (S&P) 500, where a daily reading of the yield can be obtained from your local newspaper or computer, and various other characteristics can be calculated. Then you would be 'almost' completely diversified, and you could say with some accuracy that you were not carrying any diversifiable risk. (Another important point: even if your portfolio consisted of only 20 or 30 shares, or a mutual fund with this composition, if they were properly chosen you might be able to say that you were well diversified, and carrying only an insignificant amount of diversifiable risk.)

But even though your portfolio is almost free of diversifiable risk, it still features nondiversifiable risk, because it could lose value if the economy fell apart at the seams. At a result, although you should not count on earning a return higher than the risk-free interest rate on a portfolio where all the risk can be 'diversified away', this is not the case with your portfolio if it contains nondiversifiable risk whose magnitude depends on its relationship to the macroeconomy. According to the CAPM, you should be remunerated for accepting this risk, with the remuneration taking the form of a higher return than the risk-free rate.

Calling the market return r_m, and the risk-free interest rate r_f, the risk premium on the market is $r_m - r_f$. In line with the last statement in the previous paragraph, you can expect to realise a certain percentage of this in addition to the risk-free rate if your portfolio contains nondiversifiable risk. Is there anything else to remember here? The answer is yes: if you accept risk that can be diversified, but you choose not to do so, then according to the CAPM you have no right to expect to be rewarded for your troubles. (This kind of observation leads us to sense a connection between the CAPM and the *Efficient Markets Hypothesis*, which says that 'on the average' the investor can do no better than buying a portfolio which substantially replicates the market – a so called 'no-brainer'. Both the CAPM and the Efficient Markets Hypothesis have been under widespread attack during the past few years, but for what it is worth, I prefer them to the alternatives that have been offered, especially for teaching purposes.)

Now let us look at the nondiversifiable risk associated with a single investment. What we need to begin with is a measure showing the correlation of the movements of our investment with the movements of the market. This measure is called the asset (i.e. investment) beta, or β. Ideally it would be possible to do a little statistical work in order to find out the relationship between the variability of e.g. uranium shares to the variability of the market, and if we did so we might come to the conclusion that $\beta = +0.7$: a 1 percent rise in the market would be expected, in a statistical sense, to result in a 0.7 percent rise in the price of uranium shares. This kind of reasoning means that the expected return (in percent) on the equity portion of the investment that we are discussing can be written as $r_e = r_f + \beta(r_m - r_f)$ $= r_f + 0.7(r_m - r_f)$. In other words, r_e is the (equilibrium) expected return on any shares that this firm issues. Notice also the sign of beta. Positive signs mean that the asset in question goes 'with' the market, while negative signs mean that the asset moves opposite to the market. The most often given example of the latter is gold, whose shares show a statistical tendency to move (weakly) against the market. For instance, if beta for gold shares was – 0.23, then if the market went down by 1%, gold shares would go up by 0.23%.

Suppose we continue by taking the (nominal) risk free interest rate (r_f) to be 8% (which is unrealistic, but pedagogically useful), and if the average risk premium ($r_m - r_f$) on the share market for the last 50 years or so is 6%, then our risk adjusted discount rate would be $r_e = r_f + \beta(r_m - r_f) = 0.08 + 0.7(0.06) = 0.122 = 12.2\%$: given the risk that is associated with investing in uranium shares instead of e.g. risk free government securities, the estimated cost of equity capital turns out to be 12.2%.

Conventionally, the cost of capital is a weighted average of the expected return on a firm's equity (i.e. shares), or r_e, and the interest rate on its debt, which will be called r_d; and as pointed out in Chapter 2, this weighted average is the discount rate that mainstream finance theory suggests that we use for capital budgeting: it is probably the best proxy for the opportunity cost of capital for a particular project. Using the weighted average cost of capital (WACC) to get our discount rate for capital budgeting decisions, or r^*, then we can write $r^* = r_e(E/E+L) + r_d(L/E+L)$, where E is the market value of a firms equity, and L is the market value of a firms debt. These market values, of course, might be taken as average values over a certain period.

In case the reader is curious, there seem to be flaws in all the methods that have been proposed to account for risk or uncertainty in capital budgeting, and the question must be asked as to whether there is such a thing as an optimal discount rate. But assuming that we can get a fairly accurate estimate of beta, I find something eminently acceptable about the approach that we have just discussed. If it is true that Secretary (of the US Treasury) Summers' research is still valid, and subjectivity is the key factor in the choice of discount rates, then one of the first steps in changing (i.e. improving) this very unscientific situation, is to make sure that the concepts we teach are not only logically correct, but that they they can be recognized as such by managers and others.

8.2.1 Exercises

1. Calculate the profit in the first tableau in the above section if we use 12.2% as a discount rate.
2. Calculate the profit in the second tableau in the above section if we use 10% as a discount rate, and our output is 30,000 units/year for 3 years (or a total volume of 90,000), making sure that you 'adjust' revenues and costs.

8.3 A Few Bad Nuclear Vibrations and a Comment on Option Theory

About 2 billion years ago a very large deposit of uranium ore at Okly (Gabon) experienced a self induced chain reaction that continued for hundreds of thousands of years, and created a large amount of radioactive debris. Fortunately, this waste was trapped in the deposit, and thus was 'self-stored' for the millions of years required for its radioactivity to decline to harmless levels.

Similar ambitions have been expressed for the enormous amounts of nuclear waste now accumulating over the entire globe. It too is scheduled to be locked up in the bowels of the earth where, surrounded by concrete, glass, salt, etc, it will remain undisturbed by the untrustworthy human hand, or the vagaries of nature, over the next hundred-thousand or so years. If you are an optimist, then you will believe the politicians when they say that they and their scientific support are completely in control of the situation, right down to the last low-flying neutron; while if you are a pessimist you prefer to believe the opposite. I happen to be an optimist, but at the same time I know that our politicians have failed us before, and given the opportunity they might do so again.

For that reason, they must not be given the opportunity. Let us remember that mankind has never faced a dilemma like that existing during the Cold War, where an accidental large scale armed conflict between the superpowers was always possible. What happened was that some very serious people in the two major power blocs insisted that their technicians and scientists do something sensible to prevent accidental nuclear war, and that something was in fact done on both sides of the Iron Curtain in the form of ingenious 'fail-safe' systems.

If it could be done then, it certainly could be done now. One possible solution to this problem has been proposed by Donald Carr (1978): store, and heavily guard, nuclear wastes until a new generation of reactors is available to recycle them. Eventually these reactors could come into existence – they are probably more feasible, technically, than fusion equipment – although at first it may seem too expensive to construct as many as will be needed. But, by the same token, eventually they will have to be constructed, regardless of how expensive they are.

Then there is the matter of reactor or nuclear plant safety – or the lack thereof. I seem to remember a conference at Hong Kong in which Lord Marshall – at that time a top-level member of the UK nuclear bureaucracy – assured the listeners that what he called "appalling operator mismanagement" in the (Former) Soviet Union had been rectified, and that country's nuclear program was back on track.

I certainly wonder if he knew what he was talking about, because now that Soviet reactor technology has been viewed at point-blank range, as at the comparatively well managed facility at Greifswald in the former East Germany, it is clear that many of us who live fairly close to Greifswald have been riding a lucky streak: in 1975 this installation was judged an hour away from an accident of the Chernobyl type, although the outside world did not get the details of this near disaster for 15 years.

Greifswald is now being dismantled, apparently at an enormous cost, and the process will apparently take 15 years. In addition, it has been designated a kind of 'school' for the management and technicians of other installations that will eventually have to be torn down. Let me use this occasion to suggest that it probably qualifies as a 'post-graduate' academy for nuclear engineers, because someday that plant and others now waiting for the arrival of the wrecking crew may have to be rebuilt, and it is best for all of us if they are reconstituted using a different set of blueprints.

The late professor (of physics) Hannes Alven of the University of Stockholm once said that the nuclear power industry relies on a level of perfection in which not even "acts of God are permitted". Donald Carr called this figure of speech "romantic drivel", which it almost certainly was, and went on to identify the real menace: "acts of man". This being the case, the solution to the various nuclear safety problems will have to be found on the plane of politics.

As with the CO_2 issue, some comprehensive international agreements are going to be necessary, accompanied by the threat of comprehensive economic sanctions that automatically come into effect when these agreements are violated. A great deal of money will also have to be spent on nuclear safety all over the world, although probably much less than the amount spent on the bomber and rocket fleets that came into existence during the Cold War, not to mention those long maneuvers near the Fulda Gap that many of us remember with such longing. Exactly what this kind of thing will mean in lost freedom cannot be gone into here, although I doubt if the freedom of Mr and Ms Consumer would be enhanced if the nuclear safety predicament was marginalized.

It also helps to remember that since the appearance of nuclear generated power, the world has been faced with the possibility of the widespread availability of inexpensive, homemade nuclear devices. Donald Carr alluded to an attack on the World Trade Building in New York almost 20 years before it actually happened, although in Carr's scenario a fission device was employed. Plutonium may pose a greater danger. So much so that some of us wonder why Carr became so upset when Ralph Nader called the fast breeder "maniacal". Under certain circumstances it is worse than maniacal, unless governments intend to do more to stop nuclear terrorism than they are doing at present to stop the sale and distribution of 'controlled substances' in the center of large cities, often in broad daylight, and with law enforcement people standing nearby.

That brings us to the final analytical effort of this chapter. It turns on the notion that when individuals are unsure of what their demand for something will be, and if they are risk averse, an option value may exist. For instance, they might not want to build nuclear plants because they do not want their lives complicated by the presence of nuclear waste; or, having constructed these plants, they may draw the conclusion that they are more trouble than they are worth. In both cases there are benefits and costs. In the first, they may be missing out on cheap electricity, and the waste problem may not be as troublesome as they thought; while in the second, they may come to regret the absence of the 'friendly atom' when they see how well certain other countries are getting along with it. These are the costs. On the benefit side it may turn out that unconventional, inexpensive energy sources will soon be available in very large amounts. One of the asumptions in this kind of analysis is that realistic probabilities can always be attached to outcomes, which in dealing with this particular topic tends to make it more an elegant piece of algebra than a genuine scientific exploit.

What we are working toward here is a more or less intangible example that is related to the Swedish nuclear sector, where nuclear 'disengagement' is to be commenced as soon as possible; but first we need to introduce the expression irreversibility, and this will be done with an example.

To be specific, let us ponder the ramifications of a (hypothetical) decision to tear down Notre Dame cathedral (Paris) and replace it by a drive-in fast-food restaurant. Some persons would regard this as an irreversible decision, although in the long run this is not quite true, since reasonable facsimiles could be constructed that, eventually, would be confused with the original structure by many persons. But even so we have removed, for the time being, the opportunity to admire Notre Dame's unique grandeur on a warm July night, while sitting on the patio of one of those captivating Parisian restaurants, just across the Seine.

When we put things like this, we realize that there are a great many Parisians, tourists, potential tourists, philanthropists, etc, who might be willing to band together and purchase an option which would give them the opportunity to enjoy the continued presence of Notre Dame, even if they realized that there was a large probability that they would never exercise this option. (They would never visit Paris, or if they lived in Paris would never sit on the patio of a restaurant across the Seine, etc.) Now we have returned to a language we understand: the language of option theory.

This same approach could be applied to the Swedish or German nuclear sector. If it were possible we might be willing to pay a premium to ensure that we can 'enjoy' the nuclear sector in the future, if we become convinced that this is better than being without it for one reason or another. What we want is the option to retain the nuclear sector or, in the event of merely a low degree of irreversibility, to restore it to its original condition; or even to delay the complete dismantleing of this sector, or to influence the rate at which it is to be dismantled, etc. In a utopian textbook world, a premium (in e.g. dollars) might be quoted for all these options. Perhaps we can calculate a kind of premium, but first it might be best to get something straight.

The purpose of the following exercise is to improve the reader's ability to do abstract thinking. Remember, economics is not a science in the same sense as physics or biology, but it is like a science, and the better trained its practitioners are in abstract thinking, the greater the contribution they will be capable of making to their own welfare and that of the society in which they live.

We can begin this discussion by looking at something called expected value. This sounds like a statistical concept, which it is, but it is a very simple one, and you can learn it easily. Furthermore, you need to know it, just as you needed to know the significance of an average. In fact, the expected value is the weighted average of the possible values, O, of an 'experiment', where the weights are probabilities. Consider, for example, flipping a coin or rolling a single die.

Suppose that you flip a coin once that has 1 on one side, and 2 on the other, with the understanding that your rich uncle – Uncle Moneybags – will give you, in dollars, the result of the flip. Thus you receive, with certainty, 1 or 2 dollars.

Now suppose that he tells you to flip it 100 times. You do so and get a sequence of numbers – e.g. $(O_1, O_2, O_3, \ldots, O_{100}) = (1,2,1,1,1,2,2\ldots,1)$, where, in this particular sequence, the outcome of the third roll was unity. If we sum these numbers and call the result $\hat{O}$, then we must have $100 \le \hat{O} \le 200$, where 100 means that all the numbers were unity, and 200 means that all were 2's. Actually, we expect neither of these, but something in between. Now suppose that Uncle Moneybags tells you to divide $\hat{O}$ by 100 (the number of times that you flipped the coin), and he will make you a present, in dollars, of the resulting number. If the O's added up to 147, then he gives you $1.47. Observe, the coin always landed on 1 or 2, but you receive $1.47.

We can now ask what you should have expected to receive: in other words, the expected numerical value obtained from summing the values obtained from a certain number of flips, and then dividing this sum by the number of flips. With 2 possible outcomes, and calling E(O) the expected value, you should have expected $E(O) = q_1\tilde{O}_1 + q_2\tilde{O}_2$, where $\tilde{O}_1$ and $\tilde{O}_2$ are these outcomes, and q_1 and q_2 are the probabilities of the outcomes. In the example given above, we have $\tilde{O}_1 = 1$ and $\tilde{O}_2 = 2$, and if we used an unbiased coin, $q_1 = q_2 = \frac{1}{2}$. (Remember, if you flip an unbiased coin once, you have an even chance of getting a head or a tail, and so the probability of each is $\frac{1}{2}$. Similarly, if you flip a three-sided unbiased coin once, then we would have $q_1 = q_2 = q_3 = 1/3$. Notice how the probabilities for these experiments inevitably add up to unity.) As for the calculation in the experiment above this is $E(O) = (\frac{1}{2} \times 1) + (\frac{1}{2} \times 2) = 1.5$. E(O) applies to any number of flips – although, trivially, there is no point in trying to apply it to a single flip where the outcome is 1 or 2. (Make sure that you understand the difference between the O's and the Õ's. The latter refers to the possible outcomes – two for the flip of a coin, and six for the role of a die; while the former refers to the outcome of an 'experiment'. For example, in flipping a coin 100 times, there are 100 O's, each of which has a values 1 or 2. Notice also the relationship between expected value and 'average'.)

The expected value in the calculation just above is $1.5, but after flipping 100 times you received only $1.47. What went wrong? The answer is nothing: you could have gotten any amount between 1$ and 2$; but if this were a gambling game, and you were asked to choose a number between 1 and 2, the optimum choice would have been 1.5, since theoretically you would have a better chance than the person who e.g. choose their 'favorite number' (e.g. 1.87). (Why is 1.5 the optimum choice in a situation with two flips? The possible outcomes are (1,1), (1,2), (2,1), (2,2), and the probability of each of these is ¼; but the probablity of an outcome displaying a '1' and a '2' is ½. This is your best bet, and the mathematical expectation here is 1.5.Try applying this kind of reasoning to a coin with three sides!)

Furthermore, had you flipped the coin a million times, added all the numbers you obtained, and then divided by 1,000,000, you should have ended up even closer to 1.5 than the 1.47 you obtained from the above 100 flips. Perhaps 1.498217! Of course, there is a finite (i.e. non-zero) probability that in e.g. our one million flip experiment, you landed directly on 1.5 when you carried out the division by one-million – about the same as the probability of your walking into the Mirabel in Saltzburg or the Grand Casino at Monte Carlo with a few dollars in your pocket, and breaking the bank.

Take as another example rolling a die. The expected value here, which again can be called $E(O)$, is $q_1\tilde{O}_1 + \ldots + q_6\tilde{O}_6$. The Õ's run from 1 to 6, and if this is an unbiased die, then we have $q_1 = q_2 = \ldots = q_6 = 1/6$. (Note again that the q's sum to unity.) As for our calculation, it is simple and straightforward: $E(\tilde{O}) = (1/6) \times 1 + \ldots + (1/6) \times 6 = 3.5$. If we roll the die once we get $1 \leq O_1 \leq 6$, while if we roll it a couple of million times, and divide the sum of the numbers obtain by the number of rolls, we obtain something very close to 3.5, which is our expected value.

Only one detail remains before we can utilize our new knowledge about mathematical expectation. We used the expression 'risk averse' above, and perhaps it is best to say what this means. It means, essentially, that a person would not accept a fair gamble. For instance, you would not risk the $50,000 in your bank account on the flip of an unbiased coin that rewarded you with $100,000 if it came up heads, and punished you with an empty bank account if it landed tails. (But this does not mean that would accept this bet if a win meant $105,000. You might require a much larger risk premium to participate in the gamble.) By way of contrast, even if you are risk averse, you might be willing to accept a fair gamble if the amount being risked was $5. In other words, in talking about risk aversion, some precision about the size of the gamble is helpful.

We can begin the formal analysis by assuming that we have two possible states of nature, θ_1 and θ_2, and at time T – which is in the future – we will find out which actually prevails. But at the present time we can only try to infer which of these will be experienced. The present time, t_0, is the starting point for our exercise, with an investment – or a disinvestment – taking place. Furthermore, state θ_1 could prevail at T with a (subjective) probability of q, and in this case the perceived social benefit from our venture would be B_1. On the other hand, if state θ_2 prevails (with probability $1 - q$), then the resulting social benefit will be taken as B_2. If we are talking about nuclear disengagement, or engagement, then at t_0 there is a (subjective) evaluation of the social benefit of nuclear disengagement, and this will be taken as B_0. What must be understood here is that it is not until T that the decision made at t_0 can be evaluated. If it turns out that disengagement was the correct policy, then the portion of the nuclear capacity that was not disposed of at t_0 can be done away with, and there might be some regrets that it was not eliminated earlier. However, if disengagement turns out to be the wrong move, then the remaining capacity is not reduced, and some regret will be expressed over the earlier reduction. As we will find out, the focus of the exercise is on θ_1 and B_1.

Now assume that the initial nuclear stock is unity, and the amount disposed of at t_0 is N_0, with $0 \leq N_0 \leq 1$. N_1 and N_2 then represent the commitment with regard to disengagement at T, depending upon whether the state of the world is θ_1 or θ_2. As is obvious, the starting point for our behavior at T will be $N - N_0 = 1 - N_0$. Figure 8.4 sums up the previous discussion, and in particular specifies behavior once the state of the world has become clear: if $\theta = \theta_1$, do not disengage; if $\theta = \theta_2$, continue to disengage.

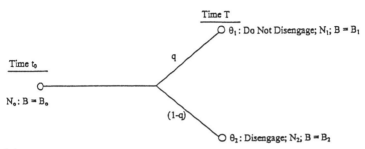

Figure 8.4

Without being concerned about reversibility or irreversibility, but remembering that we only have state θ_1 or θ_2 at time T, let us examine the expression for expected present value, assuming risk neutrality on the part of the decision maker:

$$PV = B_0 N_0 + \frac{q B_1 N_1 + (1-q) B_2 N_2}{(1+r)} \qquad (8.1)$$

The structure of this equation indicates that B_0 is the (perceived) social value per-unit of nuclear equipment being 'disengaged' at time t_0, with $B_0 > 0$. Similarly, B_1 is the perceived social value of nuclear disengagement at time T, given state θ_1. The premise is that if state θ_1 obtains, then it was a bad idea to disengage, and so a negative social value is attached to the disengagement: $B_1 < 0$, and the rational behavior at T is to cease the disengagement. In equation (8.1), N_1 becomes N_0, which is the amount already scrapped. 'r' is the discount rate.

By the same token, if θ_2 is the actual state at T, then disengagement continues. To keep things simple, it will be assumed that the remainder of the nuclear inventory (or $N - N_0 = 1 - N_0$) is terminated. In this context a positive social value must be attached to disengagement, and $B_2 > 0$. Let us now adjust equation (8.1).

$$PV = B_0 N_0 + \frac{q B_1 N_0 + (1-q) B_2 (1-N_0)}{(1+r)}$$

A slight rearrangement will give

$$PV = \left(B_0 + \frac{q B_1}{1+r} \right) N_0 + \frac{(1-q) B_2 (1-N_0)}{(1+r)} \qquad (8.2)$$

The final step in this development is simply to recognize that if state θ_1 prevails, then the second term on the right hand side of equation (8.2) drops out (since $1 - N_0$) becomes zero), and the only way that we can have a positive PV is to have:

$$B_0 > \frac{-qB_1}{(1+r)} \qquad (8.3)$$

Notice in the above the not so subtle departure from risk aversion to risk neutrality in order to obtain equation (8.3). This was to simplify the mathematics. Thus we can say that equation (8.3) understates the necessity to avoid irreversible investments – in this case disengagement – unless the value you put on disengagement is very much greater than the (discounted) mathematical expectation of things going wrong. Let's look at this a little closer. If things go wrong, you might come to think of every disengaged unit of capacity as being worth $2,000 (or, taking the sign into consideration, – $2,000). If there is an even chance of things going wrong, or $q = 0.50$, then the mathematical expectation associated with the bad outcome is the (discounted) value of $1,000. Equation (8.3) says that the value you should put on a unit of disengaged capacity at t_0 should be more than $1,000 (assuming $r = 0$). How much more? Well, if we look at that expression, one dollar should be enough, which isn't much help. More help might be provided by assuming risk aversion. Then we could say a lot more. For instance, if I were the decision maker in the example being looked at here, I might assume that it was a certainty that things would go wrong. Thus, with q equal to unity, and taking $r = 0$, I would take $B_0 = $2,000$. This is the social value that I estimate a disengaged unit of capacity must be able to provide in order to make disengagement acceptable, given that the outcome of disengagement is uncertain, and there is no way to insure against an adverse outcome.

What do I mean by 'social value'? I might mean the 'money value' of the satisfaction that I would get from knowing that the local nuclear installation has ceased to exist, as well as the pleasure in seeing the amount of electricity that it once generated being generated by a wind farm, etc. Furthermore, this should take into consideration the increased cost of electricity, should there be an increased cost. (However, I might just mean the satisfaction or 'utility', with no reference to money.)

Let's rephrase this slightly for emphasis. Before even a risk neutral decision maker forecloses his or her options by going ahead with disengagement, this person would have to be assured, at t_0, that the benefits of disengagement were worth more than $-qB_1/(1+r)$ (which is greater than zero since B_1 is negative). In these circumstances the risk would be taken, since even if things turned out to be unfavorable, the decision maker would claim that according to the economic theory that he or she had been studying, there was justification for this behavior. What that economic theory did not tell them, however, was the size of the optimal disengagement at t_0, which is a much more difficult exercise than calculating B_0.

8.4 Conclusion

The great reality of nuclear power is the apprehension aroused in all of us that something could go wrong. I certainly believe that the Swedish nuclear sector is the safest in the world, but even that may not be good enough. The latest report that I have seen about Swedish nuclear installations notes various technical shortcomings, and even more disturbing the attempt to reduce wage and salary costs (by decreasing the number of employees at these facilities).

The talk in Sweden at the present time is of ecologically durable communities, and sustainability. Unfortunately, as I see this matter, a high level of energy use – perhaps very high – will be necessary in the intial stages of establishing such communities. Consider, for example, the transportation requirements of these communities, which in theory will could involve such things as the large electric trams going from the center of Grenoble to the university, and the 'elevated line' in Sydney that goes to the Darling Harbour shopping complex, as well as fast electric trains connecting various centers of population – such as the train between Hamburg and Berlin. Personally, I am satisfied that imaginative and dedicated politicians can solve <u>any</u> problem associated with <u>any</u> aspect of energy, while at the same time making great progress in enhancing the quality of life. Whether these ladies and gentlemen will take the necessary steps to do so remains to be seen.

Chapter #9

Additions, Extensions and Final Remarks

Should the reader recall my earlier negative comments about the Hotelling hypothesis (or model), then he or she may be surprised to learn that a derivation and extension of that model is the first topic in the present chapter. The algebraic structure of the model provides a useful and not too difficult analytical exercise, and it is not completely impossible that someday it will help the reader to obtain a valuable insight into various commodity markets. At the same time let me make it absolutely clear that in the real world – what an acquaintence of mine once called "peopleland" – competition is much too stiff for students and teachers alike to become overly occupied (i.e. obsessed) with esoteric notions that may never be capable of explaining events and behavior in actual markets.

Much more useful are the other concepts taken up below. For instance, *convenience yield* is a key ingredient in discussing the term structure of oil prices, and would fit nicely into the discussion of inventories and derivatives markets that was presented in Chapters 3 and 6. Considering the career opportunities that are available in commodity derivatives markets, I think that all students of economics should make themselves aquainted with the materials in Chapter 6, as well as the other comments on derivatives markets that are found in this book.

As for real (or *management* or *strategic*) option theory (which has to do with physical investment for the most part), I am not going to claim that it is indispensible, but it is the kind of subject that you would do well to become acquainted with as soon as possible if you place any value on impressing colleagues and potential employers. The last section of the chapter is no more than a short comment, but I hope that it gives you something to think about. Energy economics is going to be an extremely important subject as the 21st century progresses – far more important than many readers of this book realize – and it is essential for persons seeking a reasonably high degree of expertise in these matters to get off on the right foot

For better or worse, the course that I promised you in the preface to this book, and in the introduction to the first chapter, has just about been delivered. Those of you who have read the book, and followed most of the arguments, are much closer now than you were to understanding some very important issues that you could encounter on a number of occasions in your personal and professional lives.

9.1 The Hotelling Hypothesis Examined Graphically

The basic Hotelling result is that the net price of a resource being bought and sold in a competitive market increases at a rate equal to the interest rate. Assuming a constant marginal cost, this net price is $p = \bar{p} - mc$, where $\bar{p}$ is the market price, and mc is the marginal cost that is associated with a given output. This rule will be derived below, and algebraically is $\Delta p/p = r$. Interested persons will be able to find this expression in some outstanding intermediate textbooks, and thus every year thousands of students throughout the world are being told that it, or one of its many extensions, contains some indispensible economic vision, when in truth its principal value is historical. At the same time I would like to note that history has a very valuable role to play in economics and finance – much more valuable than many students of economics realise. Joseph Schumpeter was one of the great economists of the 20th century, as well as a star sociologist and historian; and he made it clear to anyone who asked that, if he could 'do it all over again', history would have come first.

The way these matters will be approached in the present exposition is to show that the Hotelling rule follows from elementary capital budgeting considerations. We can begin by writing the well-known relationship for the present value of a profit stream that results from producing our energy resource over T periods:

$$PV = \sum_{i=1}^{T} \frac{\bar{p}_i q_i - c_i q_i}{(1+r)^i} = \sum_{i=1}^{T} \frac{R_i - C_i}{(1+r)^i} = Maximum \qquad (9.1)$$

In this expression $\bar{p}_i$ is the expected market price (for every period except the first, when this price can be considered known), while q_i is the profit maximising output that is consistent with that price. The discount rate (which in the Hotelling model is taken as the rate of interest) is r, while c_i is the unit production cost. R_i is then the revenue, and (in the conventional presentation) C_i is the 'non-capital' costs. Something that must be carefully pointed out is that Hotelling not only assumed constant returns to scale, but specified that marginal cost is the same as unit (variable) extraction costs. The net price can then be written $p_i = \bar{p}_i - mc_i = \bar{p}_i - c_i$ for all periods 'i'. Remember too that since we are assuming a perfectly competitive situation, $\bar{p}_i$ is obtained from the market. After the first period, it is the *expected* market price. Thus the 'program' that we are going to specify for all periods up to the terminal period may have to be revised every year as expectations change.

Now let us introduce an expression called the *marginal profit* (MP) for each period. This is $MP = mr_i - mc_i$, and in Hotelling's model this becomes $MP = \bar{p}_i - c_i$ for all periods 'i', where mr_i is the marginal revenue. Why hasn't the reader seen MP before? The answer is that in your earlier course(s) in economics, the profit maximizing condition was mr (= p for perfect competition) = mc, and so MP was zero. In fact, when MP was not zero, then production should have been increased or decreased until mr = mc. But this will not be the case where our exhaustible resource firm is concerned. Of more use to us, however, is the discounted marginal profit for each period, or $MP_i' = (\bar{p}_i - c_i)/(1+r)^i$, where the period numbering is i = (1,2,....,T), with T being the planning horizon for the producer, and indicating the final period in his program to remove Q units from the deposit. Q will show up in the graphical analysis, and appears in the discussion of scarcity, but does not enter into the derivation of the basic Hotelling result in an explicit fashion. To make the problem interesting, Q is taken as the total amount of reserves.

That brings us to the matter of marginal revenue not being equal to marginal cost when exhaustibility (or depletion) enters the picture. The reason is because we are categorically taking into account the fact that a unit of a depletable resource (such as a barrel of oil) that is sold today will not be available tomorrow, when it 'might' have realised a higher profit. Thus, in maximising the present value (of profit), extraction must be programmed in such a way as to consider the opportunity cost of earlier as compared to later production. Suppose we assume that PV is a maximum, but at the same time we have

$$\frac{MP_x}{(1+r)^x} > \frac{MP_y}{(1+r)^y} \quad (or) \quad MP_x' > MP_y'$$

Thus, if we were to take one unit of the resource and transfer it from period y to period x, we would get:

$$\Delta PV = \frac{MP_x(+1)}{(1+r)^x} - \frac{MP_y(-1)}{(1+r)^y} > 0$$

Accordingly, PV was *not* a maximum. This immediately implies that the profit maximizing condition must be:

$$MP' = \frac{MP_1}{(1+r)} = \frac{MP_2}{(1+r)^2} = ... = \frac{MP_i}{(1+r)^i} = ... = \frac{MP_T}{(1+r)^T} = \Phi \quad (9.2)$$

By way of interpretation, when PV is a maximum, the discounted marginal profit (MP′) is equal for all periods, and here is called ϕ. As should be clear from the graphical analysis that follows, it is theoretically possible to have q = 0 for some periods, namely when ϕ is greater than or equal to the maximum discounted marginal profit for that period, which is the vertical intercept of the discounted marginal profit curve. (q = 0 for that period because if it were greater than zero, then output is being sold during that period that would make a larger contribution to the *total* discounted profit if it were extracted later.)

A great deal has been made of the last observation in the learned literature, and to a limited extent it is an elegant result. The problem is that if there are fixed costs that must be paid every period, or for that matter opportunity costs are associated with closing down temporarily, and then restarting production later, then we should not expect to see q = 0 in some period, and q > 0 in the following period – although it occasionally could happen for small operations. (Small coal mines in the US have been known to function in this matter, although for the most part they too, like the major coal firms, continue to produce even in bad times, when according to our rule this coal would be saved for *expected* good times.) This by itself should be sufficient to rebuke modern interpretations of the Hotelling hypothesis although, as will be pointed out later, Harold Hotelling was fully aware of this shortcoming. It might therefore be logical for resource economists to consider repairing this deficiency, instead of expanding it, and the graphical apparatus below is a modest introductory contribution to this venture. (See in conjunction with this diagram the next set of exercises.) Put more directly, once fixed costs are brought explicitly into the analysis, then the outrageously overpublisized $\Delta p/p = r$ rule will have to be drastically overhauled.

Now let us look at a typical period such as 'i'. From equation (9.2) we get $MP_i' = \phi$, or $mr_i = mc_i + \phi(1+r)^i$. To use Salant's terminology (1995), we are now employing a more comprehensive definition of the marginal cost, which he calls the "augmented marginal cost". ϕ is defined as the present value of the opportunity cost (or scarcity rent) which is associated with producing a unit of a resource now instead of later. When is ϕ equal to zero?

It is zero when the resource is not exhausted during the chosen time horizon (T), and thus in these circumstances, analytically speaking, there is no scarcity. We also have, with $\phi = 0$, our conventional profit maximising condition: $mr_i = mc_i$.

We have now reached a point where the above exposition can be put into graphical form. Taking a situation with 3 periods, and the intention to remove Q units during that time horizon (T = 3), we might have the production program shown in Figure 9.1.

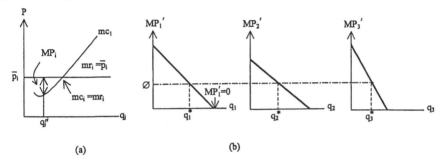

(a)   (b)

Figure 9.1

Figure 9.1a shows the situation with the firm in any period 'i', given expectations regarding costs and prices. Thus, everything in this analysis except the initial period is based on expectations, and as a result – implicitly – it is possible to raise the issue of revising expectations. In addition, the above apparatus immediately suggests a way of showing the result of increased uncertainty. This involves (as in the mainstream literature) increasing the discount factor (which in the usual exposition is the rate of interest), which in Figure 9.1b is tantamount to a downward shift of the MP' curves. What will this lead to? A higher discount factor reduces the present value of distant profits, and if correctly drawn the above (MP') curves will show production moved forward. We see this kind of result everywhere in economics: uncertainty leading to more emphasis being placed on the present as opposed to the future.

In Figure 9.1b production in the three periods is chosen so as to exhaust the particular resource. In other words, $q_1{}^* + q_2{}^* + q_3{}^* = Q$. Note also from this diagram that $MP_1' = MP_2' = MP_3' = \phi$, which is the maximum profit criterion that was (informally) derived above. With this observation behind us, and remembering that $MP_i' = MP_i/(1+r)^i$, we can derive the Hotelling rule. For example we have:

$$(mr_i - mc_i)/(1+r)^i = (mr_{i+1} - mc_{i+1})/(1+r)^{i+1} \qquad (9.3)$$

In a competitive situation we have $mr = \bar{p}$, and so the above becomes:

$$\frac{\bar{p}_i - mc_i}{(1+r)^i} = \frac{\bar{p}_{i+1} - mc_{i+1}}{(1+r)^{i+1}} \qquad (or) \qquad \frac{p_i}{(1+r)^i} = \frac{p_{i+1}}{(1+r)^{i+1}} \qquad (9.4)$$

Simple manipulation of this expression yields:

$$(p_{i+1} - p_i)/p_i = \Delta p_i/p_i = r \qquad (or) \qquad \Delta p/p = r \qquad (9.5)$$

Here it is necessary to remember that p is the *net* price $(= \bar{p} - mc)$.

The academic world is so fond of equation (9.5) that it seems a pity to do anything except to praise it, however if we look very carefully at this expression, we see that it reduces to $p_{t+1} = p_t(1 + r)$. Thus if we were talking about financial assets, rather than physical assets, we might have something that we could use outside the classroom, but I sincerely doubt if it would fit into a planning session in the executive suite at Exxon or Shell – at least, I hope that it wouldn't.

Before going to the backstop, let me say that I consider a basic flaw in the contemporary treatment of Hotelling's model to be the neglect of fixed costs. As it happens, Hotelling himself (1931) called attention to the fact that "capital intensity in developing the mine…is a source of a need for steady production," which must be one of the great overlooked observations in the entire history of economics. The question is, with "steady production", will equation (9.5) still be valid? I like to think that it won't, but at this point I leave to others the formulation of a proof or counterexample which shows that equation (9.5) ceases to be relevant when fixed-costs/steady production are brought into the analysis in a meaningful way. (See the next exercise set!)

It is also valuable to emphasize again that irreversibility dominates many of the decisions made in exploiting natural resources. Fixed costs and irreversibility are to some extent related, but the opinion here is that the term "irreversibility' was not in Hotelling's vocabulary. In any event, the longer the time horizon the more uncertain the outcome, and the more uncertain the outcome, the greater the value of procrastination – usually.

This section will be concluded with a few words about backstop technologies. Conceptually, a backstop technology is one that can be purchased at time T, for F dollars, and which – at that time – can produce one unit of some energy resource indefinitely. If the total requirements at time T is q^*, then the amount needed to put the backstop into place is Fq^*. But remember, structures and machines cannot be brought into existence overnight. The investment must begin earlier, and perhaps much earlier.

It might thus be argued that the price of the resource at the present time should be increased in order to help pay for the backstop technology. If we are in a competitive situation, with price = marginal cost, then this extra amount will raised the price above the marginal cost, as in our previous discussion; although the farther away the 'switch point' T, the smaller the increment.

If we were to go into this subject thoroughly, employing a little calculus, then we would find out that under competitive conditions, inexpensive resources should always be used first, and (under competitive conditions) will be used first, since the owners of these inexpensive resources can always undercut the price offered by the owners of more expensive resources, to include those obtained via the backstop technology. By itself, this is no great discovery; but since the *global* energy market is probably not competitive, and production costs vary widely around the world, it helps to explain the very strong position held by Middle East producers of oil, oil products, and petrochemicals, and why this position could grow much stronger as the gap increases between their production costs and those of producers in other parts of the world.

By way of completion, another piece of algebra will be added to this discussion. Professor Merton Miller is probably the leading financial economist in the world, but I happen to be worried by one of his beliefs, which is that there is no shortage of natural resources – now *or* in the foreseeable future. His writing on this topic has been limited, but among other things he and Professor Charles Upton derived the expression $V_0 = (p_0 - c_0)Q$, where V_0 is the present value of in-ground reserves, p_0 is the present spot price that results when the behavior of producers follow the Hotelling assumptions, and c_0 is the current unit extraction cost. Q is, of course, the present total amount of reserves – known with certainty.

I can derive the above expression from my equations (9.1) and (9.2), with the first step being starting the summation in (9.1) at $i = 0$, instead of $i = 1$. Going to (9.2) this gives us $p_0 - c_0 = \phi$. If we multiply both sides of this simple relationship by q_t, sum, and recognize that $\sum q_t = Q$, then we get $(p_0 - c_0)Q = \phi Q$. ϕ has been discussed above, and labled the scarcity rent, which is the same thing as the in-situ value of a unit of the resource. Thus, ϕQ is the value of known reserves Q, and is called V_0 here.

Next, the above two authors carry out an empirical investigation which indicates that the above relationship overestimates the imputed market value of oil and gas reserves in 92 percent of the 192 valuations they examine.

Accordingly, the conclusion that should be reached is that oil and gas are *not* scarce.

In a brilliant lecture given at the University of Stockholm when I was a graduate student there, Professor Karl Vind said that "In economics, empirical work can never take the place of theory." As it happens, the issue here is neither empirical work nor theory, but simple observation. *If* oil and gas markets were as Hotelling pictured them, which they are not; and *if* producers behaved in such a way as to generate the $\Delta p/p$ % rule, then Miller and Upton might have a point – but such is very definitely not the case. Fortunately, however, this research did not make an appearance in their brilliant macroeconomics textbook (1974).

Let me suggest that readers who are dissatisfied with the conclusions that I have drawn in this matter should examine Davis (1996) and Krautkraemer (1998).

9.1.1 Exercises

1. Adjust the diagram in Figure 9.1b so that production in the second period turns out to be zero! Suppose that with this adjusted diagram, the same amount, Q, was to be removed. What would this mean for extraction in the first and third periods?
2. As in the previous exercise, adjust the diagram in Figure 9.1b so that production in the second period turns out to be zero, while the total amount removed is Q. But continue the exercise by assuming that this firm has fixed costs (e.g. in the form of loans for capital equipment) that must be serviced every period. Without worrying about the mathematics, what does this mean in terms of Figure 9.1b? What does it probably mean for the Hotelling Hypothesis as represented by equation 9.5?
3. Follow the directions given above, and derive the Miller-Upton expression for the value of reserves: $(p_0 - c_0)Q = V_0$! If you know Kuhn-Tucker theory, explain why ϕ in the above discussion is the in-situ value of a resource. This result can also be obtained by taking ϕ as a Laplace multiplier.

9.2 Convenience Yield

As Cristina Caffarra (1990) pointed out in an astute criticism, analysts often pay too much attention to short-term factors in the oil market, and thus take insufficient notice of longer-term indicators. In her opinion, the "interpretation of stock levels and changes is one clear example of this". She also notes that concepts like convenience yield have deservedly become an important part of the vocabulary of the oil market.

Although I find it possible to agree with both of these observations, some reservations are in order. Where short-term price movements on the oil market are concerned, it is impossible to pay too much attention to stocks (i.e. inventories). This idea was promoted extensively in Chapter 3. In addition, convenience yield is still too much of an academic figure of speech, which is unfortunate, because if it were better understood such important topics as 'the term structure of oil prices', and 'backwardation' would be much more useful to both students of energy economics and energy professionals.

What we really need, of course, is some way to tie the short and long run together. A proposal that has been informally suggested by economists of the Institute for the Economics and Politics of Energy (of the University of Grenoble, France) turns on the elaboration and study of an oil price profile that will feature the orderly development of oil reserves, regardless of where they are found, and which implies tht oil shocks caused by events such as the Gulf Crisis would not lead to long lasting instabilities and imbalances, to include non-optimal investment practices. Calabre (1991), among others, insists that the effective operation of futures markets, and the speculation associated with them, are not destabilizing.

On the other hand, Criqui (1991) has indicated that market fundamentals become increasingly important as the quantity of oil consumed rises, which might be taken as a way of saying that conjectures about derivatives markets are becoming increasingly irrelevant. What about an optimal or 'ideal' oil price profile? Both the work of Criqui, and that of Laherrere referred to in Chapter 3, suggests that some very bad news might be waiting for oil consumers in the not too distant future, and as a result the ideal oil price profile is an idea whose time has passed, at least for oil, although it might be of interest for natural gas and coal.

That observation brings us to some remarks intended to simplify certain concepts forwarded by Caffara and Calabre, beginning with convenience yield, which in turn is generally used to explain backwardation in forward and futures markets. The backwardation that we will be concerned with here is *strong backwardation*, where futures prices are below current spot prices. Normally, we would expect this kind of disposition to be impeded by virtue of the role of storage costs: what's the point of storing something if market signals indicate that you will only be able to sell it at a lower price than you purchased it? However between February 1984 and April 1992, the 9 months futures price was strongly backwardated 77% of the time. What this means is that, unlike the situation in financial markets, there are long periods in the oil futures market when backwardation is the normal state of affairs, although there are varying degrees of backwardation. For example, during the Gulf Crisis it became extreme.

Furthermore, regulatory changes in the 1980s led to an increase in the frequency of spot transactions in the US. More important, some evidence became available that these changes led to an increase in spot price volaδility. A conclusion that everyone should be able to draw from this is that an increase in volatility increases basis risk, and as a result reduces the efficiency of a futures market for the hedging of price risk. For this reason alone, futures are a poor choice for hedging price risk in electricity and gas markets.

Now for some simple algebra, employing the symbols and concepts that were used in Chapter 6. S is the spot price; F the forward price; r is the interest rate; and 'u' is introduced to represent the unit storage cost. We have as the equilibrium relationship for a *one period* situation:

$$F_1 = S_0 (1 + r + u) \tag{9.6}$$

Notice how this works. If F_1 is greater than the right-hand-side (RHS) of this equation, then rational transactors will borrow money to buy and store the commodity, while shorting a futures contract. They will then deliver on this contract at the end of the period, which will yield an economic profit. What we are dealing with here is simple inter-temporal arbitrage.

Next suppose that F_1 < RHS. This provides another opportunity for arbitrage: sell the commodity (if you have it), and invest the proceeds at the interest rate r, while at the same time buying a futures contract. Then at the end of the period take delivery on the futures contract, which means that you end up with the commodity *and* a profit on your financial transaction. (Something else that you could do here is to borrow the commodity if you do not have it.)

However on many occasions we observe that individuals do not carry out this kind of arbitrage – or, better, the amount of arbitrage suggested by this arrangement. The reason is that they obtain a 'utility' – a *convenience yield* – from holding a given level of stocks that is more valuable than the financial gain that would be realized if they began to exchange these stocks for money. Put another way, if they considered performing this arbitrage, then they would conclude that the loss in 'convenience' – i.e. the *marginal* convenience yield – would be so large that the transaction should be deemed unfavorable, and so it would not be carried out. This implies that the above 'equilibrium' relationship must be adjusted. Thus we have:

$$F_1 = S_0(1 + r + u) - Y \tag{9.7}$$

Y is the marginal convenience yield. Are we sure of this? Well, suppose that we have very large inventories. If we add another unit to this amount, the convenience yield of these inventories remains high; but the marginal value of another unit being added to stocks is probably close to zero, or zero. Thus, with very large inventories, if F_1 < RHS in equation (9.6), and the arbitrage suggested above actually takes place, then it immediately suggests that Y → 0. Y can only be the marginal convenience yield! Figure 9.2 shows the situation, where I/C represents 'inventory coverage'. (See below!)

This expression is made more useful by assuming that $Y = yS_0$, with y > 0. The logic here is that, *ceteris paribus*, as the spot price increases, inventories become more valuable – or, what is the same thing, Y increases. Now we can write (9.7) as:

$$F_1 = S_0(1 + r + u - y) \tag{9.8}$$

Finally, it is possible to write y = y(I/C), where C is the consumption of the commodity, and I is the inventory level. Ordinarily, we should have $\partial y/\partial(I/C) < 0$, which means that when I increases relative to C ,the marginal convenience yield decreases (due to a fall in y). The most important thing about all this, however, is that this relationship is not linear. Instead it takes the following form:

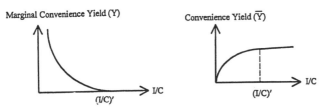

Figure 9.2: Convenience (yield) as a function of the inventory/consumption ratio.

Now let us see if we can make some fruitful observations about volatility. If the level of inventories is high, then even a comparatively large change in the level of inventories would result in only a small change in the convenience yield. Thus the change in the spread between spot and futures (or forward) prices – the basis – should be small. In Figure 9.2 we would be on the far right of the curves. The same argument could be made about a sharp change in C, given large inventories. Not only would there be a small change in the basis, but in the case of a demand shock upwards, the rise in the spot price would be moderate due to selling from inventories.

By way of contrast, the lower the inventory level, the greater the effect of changes in inventories on the spread between spot and forward or futures prices as a result of the comparatively large change in 'convenience'. In Figure 9.2 we would be on the far left of the curves, and a pronounced

240

upward escalation in demand would not be abated by a fall in inventories, but by a sharp rise in S. Thus the spread between S and F – the basis – could greatly increase. In fact, with a very low level of inventories, the change in C would not have to be especially large in order to bring about a considerable change in the basis. Increased variability in the basis translates directly to increased basis risk. Even W. David Walls (1993), a warm friend of complete deregulation, has been forced to admit that existing institutions and technologies have not been successful in preventing "coordination failure" between the spot and futures markets, which he goes on to say "allows a disparity between the spot and futures price at contract delivery".

9.2.1 Exercises

1. Read the above section very carefully, and then return to Chapter 3. Use what you learned in this section to modify, complement, or expand various topics in that chapter!
2. The diagrams in Figure 9.2 have I/C on the horizontal axis. I have seen this kind of diagram with I on that axis. Would using I instead of I/C change a great deal in the discussion.

9.3 Analytical Odds and Ends: A Statistical Observation; The Winner's Curse; and Real Options

I hope to expand the list of topics taken up in this book in future editions, while making sure that persons who use this book for self-study are given the background to deal with with certain fundamental concepts.

The first item to be examined is already well known to many readers of this book. These persons already know how the expected outcome (or return) E(X) of an investment (or 'prospect') is computed. We simply multiply the possible outcomes by their probabilities, and sum. In symbols, with n outcomes (X_i) and n probabilities (p_i), we can write:

$$\overline{X} = E(X) = \sum_{i=1}^{n} p_i X_i \quad (E = Expectations \ operator) \tag{9.9}$$

We are also aware that when considering an investment, when possible, risk should be brought into the picture. Careful readers of previous chapters probably know that we usually represent risk by the standard deviation (σ) of possible outcomes. The standard deviation is the square root of the variance (σ^2), which is calculated by:

$$VarX = \sigma^2 = \sum_{i=1}^{n} (X_i - E(X)) p_i \tag{9.10}$$

What we want are investments with a high expected return and a low risk; but most of the time we must choose between investments with low returns and low risks, and those with high expected returns but high risks. What we eventually choose usually reduces to a matter of individual taste.

Often we work with statistical diagrams of the type shown in Figure 9.3a. This is a probability density curve, and it has the well-known form of a 'normal' (or 'Bell') curve. Suppose that this curve was constructed for an investment having very many possible outcomes, which are shown on the horizontal axis, while the probabilities of occurance are on the vertical axis. Suppose also that the expected value E(X) of these outcomes is 8%, which was calculated using equation 9.9. We proceed by calculating σ, and let us say that turns out to be 10%. (Note the units for σ: they are also in percent.)

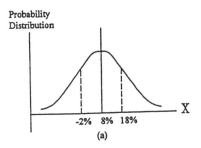
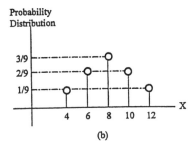

Figure 9.3

In considering the arrangement in Figure 9.3a, we know from our first course in statistics that while the best estimate of our return from this investment is 8%, the actual return is almost certain to be something else. We can feel fairly certain, however, that if this particular investment were made a large number of times, about 68% of the outcomes will be between $\bar{X} \pm \sigma = 8\% \pm 10\%$, which is the same as saying that $-2\% \le \bar{X} \le 18\%$. This is called a confidence interval. We also expect that 95% of the outcomes in a repeated experiment could be found somewhere between $\bar{X} \pm 2\sigma$.

Next, suppose that we have two '3-sided' die, and each of them has the numbers 2, 4, and 6 on the sides. If these are thrown together, the possible outcomes – and their probabilities – are shown in Figure 9.4.

	Die '2'		
	2	4	6
2	4	6	8
4	6	8	10
Die '1' 6	8	10	12

Outcome	Number	Probability
4	1	1/9
6	2	2/9
8	3	3/9
10	2	2/9
12	1	1/9

Figure 9.4

242

We can use this information to construct a probability distribution curve. This is shown in Figure 9.3b.

9.3.1 Exercises

1. Calculate $\overline{X}$ for the simultaneous cast of two 3-sided die of the type discussed above! What is the 95% confidence interval in Figure 9.3a?.
2. Calculate variance and standard deviation for the situation in Figure 9.3b. Obtain the 68% confidence interval, and interpret.
3. If the values (2,4,6) on the 3-sided die were percent, and you looked upon the prospects in Figures 9.3a and 9.3b as 'one-off' investments, which would you prefer? Why?

That brings us to the 'Winner's Curse', which is a real problem due to the growing importance of deep water operations. What we might have here is a 'sealed-bid' auction for every tract of a sea-bed gas field: bids are submitted in sealed envelopes, with the highest bidder obtaining the property. Prior to the auction, bidders can run whatever tests they desired on the potential drilling sites, and so they are immediately faced with deciding how much to invest in this phase of the project. After they have run their tests, they must attempt to estimate what information other bidders have obtained in order to formulate a sensible bidding scheme. Attempts to rig auctions by such things as cooperative agreements among bidders are strongly discouraged.

We see the winner's curse immediately: the winning bid is an amount above that which any other bidder thinks the property is worth. If you are a winner then it immediately implies that your bid was too high. Then why not bid lower, or wait until the auction was over, and then buy in the resale market? One answer here is that if the curse turns out to be a blessing, then the executive who let the property go to someone else could find himself with a pink slip in his hand being escorted to the front door by 'security'.

Governments can, of course, make these competitions more attractive to bidders by making them 'second-price' auctions, although this would hardly be rational for countries with serious budgetary problems. With this arrangement, the winner is the highest bidder, but the price paid for the property is that submitted by the second highest bidder. Professor James Quirk has cited a case in which a winner bid $111M, and the next highest bid was $6M. Unfortunately this was not a second-price affair, and in the winner's judgement, he "left $105M on the table". On the other hand, had it been a second-price auction, the government would have ended up carrying the can, and this might not have looked so good if this had been one of those situations in which the energy company had a higher annual income than the Gross National Product of the particular country.

The final technical subject in this chapter – and in this book – will be real options. (These are sometimes called 'managerial options' or 'strategic options'.) Probably more than an entire chapter in the book has been devoted to resolving some issues associated with the net present value (NPV) approach to investment; but everywhere we go these days we hear that this approach is faulty: e.g. it assumes that investments are reversible, or if irreversibility does enter the picture, then we are dealing with a now-or-never situation, and so on.

The truth is that some investments *are* reversible and some are now-or-never propositions. It may be equally true that even when it might be obvious that they are not, they must be treated (in the planning stage) as if they were, because the analytical apparatus to treat them otherwise is not available. The contention of real options theory is that there is such an apparatus, and this is also true, but whether that apparatus is as universally applicable as some real options enthusiasts like to claim remains to be seen. An example will be given below of a situation where it might be applicable to a 'real-life' situation, and that example is simple enough so that all readers of this book should get some ideal of the concept being treated, even if they remain hazy about the details.

The upshot of all this is that real options theory is interesting to study and teach, but it is inextricably involved with many of those issues which make NPV theory difficult to apply. There is no escaping the need to estimate the probability of various outcomes, and the choice of discount rates – although, as Dixit and Pindyck (1993) make clear, viewing investment as an option emphasizes risk, while deemphasizing discount rates.

The shortcomings alluded to above will not concern us here. Instead, attention will be focussed on investment – or disinvestment – as an option which 'carries' a certain value that is an opportunity cost, and which should be taken into consideration when considering large projects characterized by such things as irreversibility. For example, when evaluating some investments, discounted (expected) profits must exceed discounted (expected) costs by an amount at least equal to the aforementioned opportunity cost. We could immediately apply this kind of thinking to our Hotelling model: option theory would hardly accept the directive to shut down extractive operations simply because some concept of marginal costs – regardless of how sophisticated – exceeded an equally sophisticated concept of marginal revenue. The (opportunity) costs that are associated with closing down and then reopening again later could be too large, since it might mean an irreversible loss of tangible capital, as well as intangible capital in the form of the specialized skills of those employees who dispersed to other occupations and localities. On the other hand, continuing to operate – perhaps at a reduced output – might make it unnecessary to undergo the costly process of reorganizing production later.

A simple example can now be presented. Suppose that it is suggested that a large power plant, burning garbage, should be constructed near the wonderful ski resort of Åre (in Northern Sweden). Everyone is not enthusiastic about this idea. They say that while a successful plant might realize a NPV of ten million dollars (= $10M), an unsuccessful plant would mean a negative NPV of $–10M. This latter bad news is countered by the suggestion that, initially, a small pilot plant costing one million dollars should be constructed, and tested for a year.

The opponents of the plant also have an argument against this. They say that the expected NPV of the large plant is zero – if it is built, and *if* the probability of its success is ½ (which everyone agrees on). This (expected) NPV is obtained from the following relationship: $E(V) = 10p + (-10)(1-p) = 10 \times 0.5 + (-10) \times 0.5 = 0$. A pertinent question is then raised: why should $1M be spent for the pilot plant, when the expected value of the major investment is zero.

Real option theory does not like this way of looking at a problem of this type. Instead, explicit attention should be paid to the possibility of changing course in the future. With more information available as a result of constructing the pilot plant, management may decide to start the project, or abandon it, or delay its start, or to make it larger. Returning to our example, the diagram below provides a good view of the issues that are being addressed.

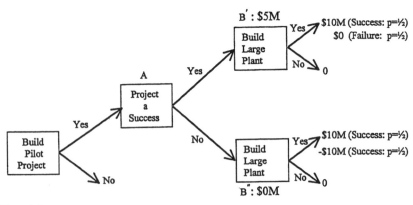

Figure 9.5

The logic here is not difficult to follow. If the pilot plant is a success, the probability is taken as ½ that the large plant will be successful, and yield a NPV of $10M; and ½ that it will be a failure, and the NPV will only be zero.

On the other hand, if the large plant is constructed when the pilot is a failure, a success for the large plant still means a NPV of $10M, but a failure now means −$10M. The reader may object to the probabilities used in this part of the example, and even the expected NPVs, but this is his or her privilege. After all, some promoters of real options theory seem to think that it eliminates the subjective elements associated with capital budgeting, but this is not true.

Notice the situation at the B-terminals! At B' the expected value is +$5M, and at B" it is zero! (Why?) Thus if we have to choose between B' and B", given that the pilot installation is a success, then we choose B'. Furthermore, we know from backward induction that the 'value' of terminal A is the maximum of B' and B", and this is B'. Going back a step further brings us to the point where the initial decision must be made, and looked at from the real options point of view, that decision should be to do the pilot project. The expected NPV obtained from respecifying the investment opportunity to include the advantages associated with delaying initiation of the large investment, turns out to be +$5M (instead of $0, as previously attained), and the cost of the pilot project is only $1M.

The upshot of all this is that by choosing to do the pilot project, we are buying an option to build (or buy) the larger plant. The $1M is the premium of the option, and the cost of the large plant (which is not given) is the exercise price. Note how this real option approach differs from the quasi 'game theory' approach in the previous paragraph. What is being said is that with an expected NPV for the large plant of +$5M, it makes sense to buy an option having a price (i.e. premium) of $1M, since the chances are that it will be worthwhile to exercise the option. This is precisely the kind of situation that option buyers want to find themselves in.

What about calculating the correct price of such an option using a version of the Black-Scholes formula. I have heard of this being done – probably in some class or seminar room – but in point of truth I am not impressed. As far as I am concerned, obtaining the variance of possible yields on a physical investment is not the straight-forward exercise that it often appears to be in the better textbooks, and the heavy use of subjective probabilities does not strike me as a particularly scientific departure. The thrilling cosmos of quantitative economic theory is probably a better place with real options theory than without it, however calculating the "correct" option premium for many prospects is something that many of us should not be trusted to do.

9.4   Conclusion: From the 20th to the 21st century

Everything that was is no more
Everything that will be is not yet.
— Alfred de Mosset

Dennis Gabor, who won a Nobel Prize in physics in 1971, once spoke of "forming the New Man who can be at peace with himself and his world". Assuming that he also meant the New Woman, this makes a great deal of sense, although some question remains as to how this forming is to be done. Gabor, however, supplied a partial answer when he said that millions of persons can contribute if they increase their education; and here I recall a casual remark my a member of my platoon when he spoke of returning to school after he left the army: "The more I know, the better I feel".

At the end of the 20th century, we are hearing a great deal more about how science and technology will supply us with the energy resources that we will need in the 21st century. Somewhere behind this almost desperate insistence is the growing knowledge that in a world where a hundred million new mouths are added every year, and where great impersonal trends will continue to be the rule, some resources could already be scarce. Scarce how? Scarce in the sense that there may not be enough of them to provide some of us with a future as satisfactory as our past – in material, social, and/or psychological terms.

At some point in the 21st century it will be discovered that in the richer countries, quantitative growth will have to slow down, and qualitative growth will have to accelerate. Many observers feel that when the time comes for that trade-off, considerably less energy will be required. The opinion here is that the opposite is true. The energy cost of high quality communities will be as large, or even larger, than what we see in the Scandinavian countries – where, until recently, a decent quality of life was considered the unimpeachable right of *every* inhabitant. In the 21st century, electricity will continue to dominate the center of the stage, and this will be true in the long as well as the short run. In the very long run the source of this electricity will most likely be wind, sun, biofuels, and so on, but that will be the very long run. Before that we will likely have to experience what Ken-Ichi Matsui calls the Seventh Energy Revolution, which he pictures as being based on nuclear energy.

Some readers are undoubtedly wondering when is the optimal time to begin this Seventh Energy Revolution – if there is such a thing as an optimal time for the initiation of this project. The real option theory that we discussed in the previous section suggests that we should wait as long as possible, and that we should not hesitate to continue to pay a premium in the form of very high expenditures for alternative energies before we move to expand or contract our nuclear sectors. In conventional real options theory, these premiums would be regarded as Research and Development (R&D) costs whose purpose is to provide information about the optimal configuration of future energy structures.

What about vehicles, which as a producer of Greenhouse gases are in a class by themselves. James A. Dunn of Rutgers university has made what I think is the most sensible comment on this problem. He understands that people do not want restrictions on their ability to use their cars, and there are probably only a few democracies in the world where they will tolerate such restrictions. However at the same time that they want new automobiles that will meet their individual need for mobility, they have expressed a "collective need for a cleaner environment." This points directly to improving the technology of the auto itself, even if it means that automobile owners must help to pay for this improvement. (The amount that they would have to pay, incidentally, might be extremely small if the research on this technology were centralized instead of devolving on a large number of automobile firms and universities.)

As already noted, technology is being counted on to keep oil inexpensive. Professor Morris Adelman has assured us in his many writings that the world oil industry will never run out of oil – "not in 10,000 years," he claims. As he correctly points out, every mineral industry is engaged in a perpetual tug-of-war between diminishing returns and increasing knowledge. "From time to time," he has said, "we win a few, lose a few, but overall humanity has won big – so far." My comment on that, in case a comment is helpful, is that in 10,000 years, or for that matter 20,000, there should be sufficient oil to fuel the lamps of China or Beverly Hills; but as for there being enough to keep your Volvo or Cadillac in the fast lane, well, that's not so certain. An observation by a friend of technology, the present Chairman of the Board of Governors of the Federal Reserve System (US), is also germane. According to Alan Greenspan, "Forecasting technology is a dauntless exercise, and we have no reason to believe that we will be better at it in the future than in the past." If this is true, and I believe that it is, then all upbeat forecasts should be heavily discounted.

As the 20th century grinds to an end, and the 21st looms before us, delegates to the major energy conferences have had it made clear to them that 'deregulation' is almost as important a topic as the future of the oil price. However, as the European Union of the Natural Gas Industry (Eurogas) has pointed out, the deregulation crusade often reduces to the advocating of competition for the sake of competition. Moreover, in the country that is providing the widely recognized "intellectual example" of electricity deregulation, the UK, observers like Professor Catherine Price and

248

Ms Yvonne Constance have shown that as a result of deregulation, many households are worse off. Similarly, the director of the Gas Consumers Council has said that pricing policies under deregulation "shows a disregard for the spirit of competition", although what she should have said was regulators, deregulators, and reregulators are just about helpless against very large oligopolies that know how to push the rules to the limit. The only way to inaugurate the "spirit of competition" is to change rules that do not work, but unfortunately nobody seems to know any longer how to do this or what to put in their place.

Having been fortunate enough to live in Sweden during the first half of the 1980s, when a majority of the country's residents seemed to believe that creative intelligence and good-will could solve most economic and social problems, I find it impossible to avoid insisting that such could again be the situation if we desired it badly enough. We still have the chance to create the energy and environmental future that most of us deserve; and certainly the ongoing deterioration of the macroeconomy does not have to be accepted – and this is true in many countries.

How long we will continue to have this chance remains to be seen. In February, 1998, the lights went out in the central district of Auckland (New Zealand), and in one of the most modern districts of one of the most modern cities in the world, they stayed out for more than a month. As far as I am concerned, if the present attempt to transmogrify energy policy to competition policy is not modified, then this kind of thing is going to repeatedly happen in the larger urban areas of many Third World countries, where utility managers will be faced with matching supply to a growing demand while, at the same time, keeping standby capacity at a minimum. And not just developing countries! As pointed out in this book, the deregulation crusade began in the United States in the 1980s when, it was claimed, the supply of power had outrun demand, and that in the future large generating facilities were uneconomical. Now, at the beginning of the 21st century, the United States has emerged as the largest buyer of generating equipment in the world. In fact, according to the *Financial Times* (June 10, 1999), the US has "rescued" the international power supply industry; and in a later article in that publication (July 2, 1999), it was pointed out that "the surge in demand in North America has been triggered by power shortages in some parts of the US over the past year, caused partly by under-investment in power stations during the 1990s." This under-investment almost certainly included, in some cases, building power stations that were too small.

By way of concluding this book, let me say that I think that almost all readers will find the technical analyses both timely and relevant. I also feel certain that many readers will agree with most of the 'opinions' that have been offered, and I am equally certain that many others will disagree with just about everything that I have said. But, regardless of where the individual reader stands, it might be useful to remember that what the great international grandmaster, Savielly Tartakover, has said of chess applies equally well to love, war, politics, and economics: "The blunders are all there, just waiting to be made." Even more important, *after* being made, most of them remain to be made again.

Appendix

A.1 Extra Questions

1. An enormous flaring of natural gas has taken place. Why? Does this make any economic sense at all?

2. Suppose that you are 20 years old, and decide to work until you are 50. After that you will take a nice 5 year vacation, traveling all over the world. According to your calculations you will need $50,000 for each of these 10 years. Assuming that you save the same amount every year, how much would this annual savings have to be in order to make this plan work out?

3. An oil lease is purchased for $1,000,000, and it is estimated that thereplace in several parts of the world. Why? Does this make economic sense? P-N Giraud does not believe that the price of coal should be indexed to the price of oil. What do you think about this? How do you feel about indexing the price of natural gas to the price of oil.

4. Some persons say that if the development costs of new nuclear equipment are put into the price of that equipment, then electricity with a nuclear background will be more expensive than other sources of electricity. Comment! How do you feel about scrapping a twenty year old nuclear facility and replacing it by a wind farm with the same power rating? What about a forty year old installation?

5. Suppose that you borrow $10,000 for 2 years, at an annual interest rate of 10%, and this debt must be serviced twice a year. What is the effective rate of interest? Hint: more than 10%!

6. What are the units in which Equation 1.1 are expressed are 400,000 barrels of oil on the land involved. It costs $100,000/year exclusive of

depletion charges to obtain the 80,000 barrels that you plan to market each year. The effective tax rate is 40%, and you estimate that the oil can be sold for $20/b. If the interest rate is 10%, what is the present value of the after-tax cash flow for this lease if unit depletion is used?

7. Discuss the concept equilibrium, and tell why it is important! Why do we call the economics we study equilibrium economics?

8. In equation (3.5) it is said that the 'Betas' obviously add to unity. Prove this!

9. In the conclusion to Chapter 3, I say that "the kind of price systems that we take such pleasure in discussing in microeconomic theory, on any level, cannot deal with this kind of phenomenon." What kind of phenomenon? Discuss!

10. Look at Figure 3.3, and see if you can formulate an equation having to do with how price moves! (Hint: consider Equation (1.3))!

11. What is a 'swing' producer?

12. Examine problem 4 in the exercises at the end of Section 4.2. Adjust the equation in that problem so that it signifies constant returns to scale. Now explain to your class, with a numerical example, why you think that it is constant returns to scale! Can you provide some kind of algebraic demonstration that will reveal it to be constant returns to scale?

13. What is optimal about point Z in Figure 4.7d? What is the (statistical) probability of a demand curve going through that point?

14. Assume that the peak and off-peak demand curves in Figure 4-8a are parallel to each other. Write out and discuss equations for them. How would the demand curve 'D' look?

15. In solving the numerical example associated with Equation (4.20), and Figures 4.8a and 4.8b, it was said that q=40 did not function with the constraint $1000 \leq q \leq 1400$, when $\beta = 4.5$! Explain!

16. What is the core, and what is special about it? In our discussion of the core in Chapter 5, do you think that there are outcomes in which players do not end up in the core? Why might this come about.

17. One expression that has not been used very much in this book is 'external diseconomies'. Considering the topics that have been taken up, where do you think that it should have been used? Explain in detail!

18. Explain backward induction in a 'general' sort of way – i.e. using words as much as possible (as compared to symbols or numbers!)

19. Figure 6.1a shows an aggregate demand curve for speculators. Draw individual demand curves for 2 different speculators, and then combine them to get the aggregate demand curve. Show the equilibrium, and discuss.

20. Show exactly how we moved from equation (6.2) to Equation (6.3)!

21. In the film 'Trading Places', Dan Ackroyd and Eddie Murphy end their adventures on the commodities futures market buying contracts for orange juice. What went on earlier? (Observe, buying these contracts apparently ended their contact with the exchange. After they made these purchases they left to enjoy the opportunities available in this world for rich young men!)

22. Would you rather be a 'writer' (i.e. seller) of American or European options?

23. In viewing Figure 6.4, we might say that there is a 50% probability that an option is at-the-money or in-the-money; and a 50% probability that it is at-the-money or out-of-the-money. What is the probability that it is exactly at-the-money?

24. A swaps market is somewhat like a futures market without speculators. Ceteris paribus, does that make it more or less attractive.

25. Theoretically, what are the opportunities for speculation on a swaps market?

26. Is the clearing house in a futures market the same as an intermediary in a swaps market.

27. M. King Hubbert worked with logistic curves. Explain how they look and why he probably became interested in them .

28. When generating electricity, oil seems to be the energy resource of last resort. But still a great deal of it occasionally used in Europe and the United States. Why?

29. How would you – the reader – handle the stranded costs problem?

30. Shareholders in the UK became very annoyed with the 'Electricity Regulator' on one occasion because of the value of X that he chose for equation 7.8. Can you tell why they were dissatisfied?

31. What are 'brownouts'?

32. There are times when buyers prefer short-term contracts for oil or uranium, and times when they prefer long term contracts. Comment!

33. What is the difference between *ex-ante* and *ex-post*?

34. In diversifying a portfolio, assets with a certain kind of 'beta' can be very useful. What kind of betas are we thinking about here?

35. The expected value $E(X)$ if we roll one die is 3.5. What is the expected value if we roll two at the same time? Suppose we roll a die that has '2' on all the faces 1000 times. What is the mathematical expectation and variance of the outcome?

36. Discuss the significance of Equation (8.3)!

37. In Hotelling's world, what happens when we have $\Delta p/p < r$ for a depletable resource? When is the scarcity rent of a resource zero?

38. Have you ever heard the expression 'the term structure of the oil price'? Explain if you have!

39. How is it (scientifically) possible to use the expression 'there is nothing wrong with using 2000 barrels of electricity to obtain 1000 barrels of natural gas'?

40. Seismic 'miracles' are making it possible to find oil and gas that would have been impossible to locate just a few years ago. But these miracles may be bringing some bad news too. Explain!

41. What is the point in using a future to hedge a transaction when you can use an option? After all, with an option you can make a profit when you hedge!

A.2   Answers and partial answers to selected exercises.

Why selected exercises and not all exercises? The reason is that most of the exercises in this book are not difficult. The reason for doing them is to get further into the rhythm of the subject, and if you do at least half of them you have every reason to feel good about yourself. Of course, if you really become interested in the subject , then you should try to do most of the problems, while remembering a famous adage from chess: the mistakes are right there on the board, just waiting to be made!   The answers are given by chapter and section number: e.g. (2.2.1/3) refers to chapter 2, section 2, problem 3!

1. (1.2.1/3) Using the dimensional scheme at the end of Section 1.2, you can calculate that the price is 4.2 cents/kWh. (Show this calculation!) In comparing this result to that in the example taking coal as the fuel, we see that the fuel cost of oil is much higher. The reason that applies here is that the $/Btu cost is much higher for oil than for coal. But remember, we are talking here about the fuel cost, and not the cost of electricity. To get the latter, the capital cost must be taken into consideration.

2. (1.3.1/4) Language is probably more important than formulae where this exercise is concerned. (a) 1 Btu raises the temperature of 1 pound of water by 1 degree Fahrenheit. (b) 2.2 Btu raises the temperature of 1 kilogram of water by 1 degree Fahrenheit (since 1 kilogram = 2.2 pounds), which is the same as saying that the temperature is raised by 0.5556 degrees centigrade, (since $\Delta C/\Delta F = 0.5556$)! (c) From (b) we get immediately that it takes 3.9597 Btu to raise the temperature 1 degree Celsius. It also happens that 3.9597 Btu is almost 1 kcal, and if we allow for the many 'rounding' errors in this chapter (which were almost unavoidable), we have every right to say that 1 kcal raises the temperature of 1 kilogram of water by 1 degree Celsius.

3. (2.1.1/5) Here is a partial answer. Consider $\log_{10}100$. As you remember from secondary school math, or from your pocket calculator, the log to

the base 10 of 100 is 2: $\log_{10}100 = 2$. We also notice that $10^2 = 100$, which leads us to suspect, correctly, that the log of 100 is the number to which 10 is raised – i.e. 2 – in order to get 100. Now you try $\log_{10}1000$! Continuing, if we have $A = e^x$, then the logarithm to the base 'e' of A (i.e. Ln A) is x. If we desire we can proceed by writing Ln $A = x$ Ln $e = x$, since the natural logarithm (Ln) of 'e' is unity. ($e \approx 2.817$, of course.) What about the $\log_{10}X$ when we have the simple exponential form $X = 10^y$? Here we have $\log_{10}X = y \log_{10}10 = y$, since $\log_{10}10 = 1$. And so on! Notice how we use log and Ln in this discussion. The second, 'Ln' refers to the natural logarithm, or the logarithm to the base 'e', while the first usually refers to logarithms to the base 10, but can refer to logarithms to any base except 'e'. Finally, we could have written Ln as ln, where ln is generally preferable.

4. (2.2.1/2) We begin by writing the expression for PV when we have an 'infinite' number of periods: $PV = V_1/(1+r) + V_2/(1+r)^2 + \ldots\ldots +$ $V_i/(1+r)^i + \ldots\ldots$ When $r = 0$ we have $PV = V_1 + V_2 + V_3 + \ldots + V_i + \ldots\ldots$ So we get PV $\rightarrow\infty$. Why not PV $= \infty$? Because infinity is a direction and not a number!

5. (2.3.1/3) Offhand, a good guess would be 1000r, unless she decided that she would pay for it in T years. Then her annual payments would be determined by the annuity formula, which you can write down, inserting 1000 in the right place.

6. (2.3.1/5) This problem is worth thinking about because the 6% is a yearly rate, but you make payments every 4 months – i.e. 3 times a year. The total number of amortization periods is thus $10 \times 3 = 30$, and the rate of interest at which the calculation will take place is $6/3 = 2\%$. Using the annuity formula you get as payments per period $11,162$, which you may proceed to check! Note that in the 30 periods you end up paying $11,162 \times 30 = \$334,874$. Now, suppose that you only serviced this loan once – at the end of 10 years. Then you would have as your payment $250,000(1+0.06)^{10} = \$447,712$ How do we explain this? The answer is that if you pay the lender $11,162 every 4 months, she can immediately invest it at the going rate of interest, which – in the calculation – is assumed to be 6% if taken as an annual rate. To make sure that you understand this point, and those discussed below, try these manipulations with $10,000 borrowed for 2 years at 10%, with repayments taking place every 6 months. But, there is still another matter that must be considered with regard to the above. The given rate of interest is 6%, and as we saw, if you repaid the loan at the end of 10 years, with compounding once a year, you paid $447,712: but if compounding took place every 4 months, then you paid $250,000(1+0.02)^{30} = \$452,840$. Thus, the interest rate applying to your loan is actually more than 6%. Let us take a more

dramatic, but easier example to illustrate this point. Suppose that we have $r = 15\%$, and compounding will take place every month during 1 year (m = 12, and n = 1). After one year the return on a dollar is \$1.1608 ($\approx$\$1.161) which you should check. For the year in question the effective interest rate is $1.161 - 1 = 0.161 = 16.1\%$. Thus, if you borrow a dollar at 15%, and must amortize it every month (at 9 cents), then the actual interest rate that you are paying is just over 16%. One final operation might be useful for the reader. If an amount L is invested for n years at an interest rate r, compounded m times per annum, we have for the final value (FV): $FV = L[1 + (r/m)]^{mn}$. Now suppose that L is compounded continuously at r'. FV is now: $FV = Le^{r'n}$. We continue by converting a rate where the compounding frequency is m times per year to a continuously compounded rate, and vice versa. To do this we set the final values equal to each other, or $Le^{r'n} = L[1+(r/m)]^{mn}$. From this we get immediately: $r' = m \, Ln[(1+(r/m)]$ and $r = m(e^{r'/m} - 1)$. For instance, if you are quoted a loan at 10%, with continuous compounding, but your payments are made every 6 months (m=2), the annual rate is $2(e^{10/2} - 1) = 2(e^{0.05} - 1) = 10.25 = r$. Thus, on a loan of \$1000, you pay \$512.5 every six months. In the above algebra, notice how the n dropped out.

7. (2.4.1/2)Depreciation methods are usually selected with regard to the effect on income taxes; the effect on a firm's profit and loss statement – which shows the revenues and expenses of a firm for a stated period; and the ease of calculation. The straight line method fits the last criterion perfectly, but obviously it could happen that the owners and/or managers of a firm might benefit if a more complicated accounting method was used – in particular one that featured faster depreciation write-offs than straight-line depreciation, and thus the payment of taxes could be postponed to a later date, which on occasion can be a very good thing. It may be so that in the application of straight line depreciation, the rule is that the cost of an asset minus its estimated salvage value is deducted as an expense in equal annual installments over the useful life of the asset. This way of looking at things does not seem completey reasonable because who, today, is so gifted that they can guess what a salvage value will be in 10 years; and who can predict what the useful life of an asset will turn out to be? It could be argued that it is better to arbitrarily specify a number of years for the write-off period, and then let the owner of the asset pay taxes on the amount received when the asset is sold.

8. (2.4.1/2) Depreciation methods are usually selected with regard to the effect on income taxes; the effect on a firm's profit and loss statement – which shows the revenues and expenses of a firm for a stated period; and the ease of calculation. The straight line method fits the last criterion perfectly, but obviously, it could happen that the owners and/or managers

of a firm might benefit if a more complicated accounting method was used. For instance, the double declining balance and sum-of-the-digits methods are highly favored in some cases because they feature faster depreciation write-offs than straight line depreciation, and thus the payment of taxes is postponed to a later date, which on occasion can be a very good thing.It may be so that in the application of straight-line depreciation, the rule is that the cost of an asset *less* its estimated salvage value is deducted as an expense in equal annual installments over the useful life of the asset. This way of looking at things does not seem completely reasonable to me because who, today, is so gifted that they can guess what a salvage value will be in 10 years; and who can predict what the useful life of an asset will turn out to be? It could be argued that it is better to arbitrarily specify a number of years for the write-off period, and then let the owner of the asset pay taxes on the amount received when the asset is sold.

9. (3.2.1/4) Merely substitute $R_{t-1}(1+g)$ for R_t in equation 3.7, and then show the people something!

10. (3.4.1/2) You will have to go long in actuals (or 'the underlying'), and so you must do something in case the price rises. Suppose you 'do something', and after you do it the price suddenly starts rising. What is your next move?

11. (4.(2.1/5) After substituting $P_2 = \bar{P}_0$, and $P_0 = \bar{P}_0$ we get (a) $Y = a_1D + a_2P_1 + a_3 \bar{P}_0$, and $H = b_1P_1 + b_2 \bar{P}_0$ implies that $P_1 = (H - b_2 \bar{P}_0)/b_1$. $Y = a_1D + a_2(H - b_2 \bar{P}_0)/b_1 + (a_2/b_1)H$. (b) I might find it possible to argue that b_2 should not be positive. It the inlet pressure (P_0) is high, then less H might be required to get a high outlet pressure (P_1). (c) Take e.g. all a's and b's equal to unity. Then for the last equation above we have $Y = D + H$. The isoquants here are straight lines, or:

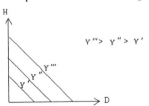

Figure A-1

In this particular case we have constant returns to scale. For instance, if we have $Y = D + H$, and then increase both D' and H' by a factor 'x', we get $Y'' = xD' + xH' = x(D' + H') = xY'$. Y is also increased by the factor 'x', which signifies constant returns to scale.

258

12. (4.4.1/6) We write for our production function:

$$q = \min[(K/a), (L/b)] = \min[(13/1), (30/2)] = \min (13,15) = 13$$

To produce 13 units requires 26 units of L, and so 4 are in excess. Thus the marginal Product of L is zero.

13. (5.3.1/1) One possibility is merely to draw a straight line from the origin in Figure 5.3a to (G_o, G_p), wherever that is, and then extend it to the core.

14. (5.3.1/5) Hint: this matter has been brought up in the section, and actually has to do with obtaining a certain degree of environmental/ecological conformity in a world characterized by 'spillovers'.

15. (6.3.1/1)The aggregate $D_h - S_h$ curve might have a vertical section. Where? Explain why!

16. (6.3.1/2) One possibility is that if jet-fuel is purchased for dollars, a jet-fuel consumer in e.g. Europe might buy dollar futures with e.g. francs. Another possibility is to hedge jet-fuel with ____!

17. (6.3.1/4) We might have the following arrangement, where the curve on the left becomes the curve on the right. You can provide the explanation!

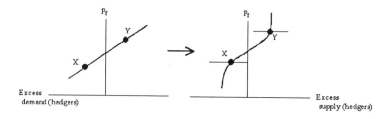

Figure A-2

18. (6.3.1/7)Hint: you compare the price on forward contracts with your beliefs about the spot price in the future.

19. (6.4.1/1) This question, and the next three, are given almost every term to the students in my course in international financial markets, which they take immediately after reading an introductory book in economics. They inevitably answer these questions correctly, and so you should be able to do the same!

20. (6.4.1/5) In theory they can be unlimited! Given the symmetry of the profit expressions for put and call options, why should this be so? Try to answer in no more than two sentences!

21. (6.5.1/4) A more precise, and easier, question would be concerned with the part that the difference between S and E plays?
22. (7.1.1/2) Conventional supply curves, when they can be drawn, reflect short run marginal costs. But capital costs must come into the picture somewhere! What did you learn about the difference between short and long run curves in your earlier course (or courses)?
23. (7.3.1/2) The cost lines intersect at 8760 hours. Prove this by, for example, working with expressions for the cost of an individual technique at 8760 hours, or $C = F + 8760V$. In this expression is 8760 the slope of the cost line?
24. (9.1.1/2) Without taking fixed costs into consideration we have the scheme on the left. When we take these costs into consideration we might have the arrangement on the right. In order to pay fixed costs, it may be necessary to deviate completely from a Hotelling pricing framework, depending on such things as the number of periods, the total amount to be extracted, etc.

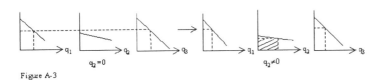

Figure A-3

You complete this diagram – labeling every axis, etc. In the scheme on the right, q_2 must be positive in order to pay fixed costs! Discuss!

A.3 GLOSSARY

This glossary is an important part of the present book, and readers should treat it as such. Unfortunately I am sure that I left out some terms that should have been included, while including some that should have been left out, but even so energy economics students and energy professionals should find it helpful.

Adverse selection: A situation characterized by persons knowing more about the risks they face than others, and can use this knowledge in e.g trading with less informed persons. This is particularly brought out in the complications associated with the setting of insurance premiums.

Amortized loan: A loan with payments of equal size that consist of both principal and interest.

API gravity: A rough measure of the density of an oil. The higher the API gravity (or number) the lower the density.

Arbitrage: 'Riskless' profit, usually gained by buying cheap and selling dear.

Asset: Something of value: share, bond, property, machine, etc.

Asmmetric information: A situation in which buyer and seller have different information about a potential transaction or commercial situation.

Backwardation: A term structure of prices in which the present (or spot) price is higher

than forward for futures prices.

Backstop (technology): The substitute for a depletable resource which can be produced at a known cost once that resource is exhausted.

Beta: A measure of the systematic risk of a given asset relative to the market as a whole, where systematic risk is nondiversifiable risk: risk that cannot be eliminated by holding a well-diversified portfolio.

Book value (of an asset): The original value of an asset less its accumulated depreciation at any point in time. Otherwise, or also, the value of a firm's assets as shown in its financial reports and the firm's 'books'.

Breeder reactor: A reactor in which more plutonium is created than fissionable fuel is consumed, and as a result most of the natural uranium input can be transformed to plutonium and used as fuel.

Bundling: A pricing strategy that involves selling two or more goods as a package.

Capital asset pricing model (CAPM): An economic model that expresses the (equilibrium) relationship between (expected) return and systematic (i.e. nondiversifiable) risk for an asset or a portfolio. This model says that the risk premium for an investment depends on the correlation of the investment's return with the return on the entire stock market.

'Caps' and 'collars': Caps guarantee that a price will never go above some level; while collars guarantee that some price will never go above one level or below another level. In this book mentioned in connection with swaps.

Choke price: The price at which the users of a good – usually a depletable resource – will switch entirely to the use of a substitute good.

Clean Air Act: Signed into law in the US in 1970. Caused a shift to more expensive fuels since, conventionally, clean fuel has always been relatively high priced.

Clean coal: Coal that has been crushed and washed, and then much of its waste is removed.

Coal gasification: This involves heating the coal (as in coal distillation), and utilizing the reaction of the solid residue. The distillation phase releases a certain amount of gas, while the entire activity produces a gas that is essentially a mixture of hydrogen and carbon monoxide.

Common property resource: A resource, such as air, to which everyone has access. (Common Property: property without a well defined owner.)

Contango: A term structure of (e.g. commodity) prices in which forward or futures prices are higher than the present (or spot) price(s).

Cooperative game: A game in which players can negotiate binding contracts that allow them to plan and employ joint strategies.

Contestable markets: Markets where firms can enter or exit freely without incurring sunk costs. (A cost that can no longer be avoided. See below!)

Customer charge: (See usage charge.)

Depreciation: Literally, the reduction in value of an asset due to use and/or the passage of time. However at the same time there can be something called 'depreciation for tax purposes' that can be an entirely different thing from actual or physical depreciation that, in any case, is difficult to measure.

Derivatives markets: Markets involving the trading of 'paper' assets such as futures and options, but also swaps and various 'structured' commodity arrangements such as caps, collars, etc.

Discount rate: A rate used to compare the value of dollars received in the future to the value of a dollar received at the present time. Not necessarily the interest rate.

Dominant strategy: A strategy that a player would want to use, regardless of the behavior of other players.

Durable good: A good that can provide services over a long time. For example, a car or a refrigerator.

Economic profit: The difference between revenues and costs, including any opportunity costs. This latter observation helps to distinguish economic profit from accounting profit.

Economic rent: The difference between the payments made to a factor of production, and the minimum amount that must be spent to obtain the use of the factor.

Efficiency: Efficiency has many facets. If we are thinking about generating electricity, then it involves how much of the energy in e.g a primary energy source (such as a fossil fuel) can be transformed to mechanical work or its equivalent (e.g illumination). But in general, in economics, we have other ways of speaking of efficiency. For example, productive efficiency, meaning that goods and services are produced at the lowest feasible cost; and allocative efficiency, meaning that goods and services go to – i.e. ar allocated to – those consumers who value those goods and services most highly, as indicated by their willingness to pay for them. Clearly, we have a welfare economics problem with this last definition.

Emissions Standard: A legal limit on how much of a pollutant may be emitted.

Energy: In all forms except heat, energy is the ability to do work. Energy and work are measured in the same units: the foot-pound, the joule, the kilocalorie (kcal), the Btu, etc.

Entropy: The degradation of energy (through, for example, a decline in its temperature), involves an increase in its entropy – i.e. of energy not available to do useful work. According to Professor Francis Sears, entropy is the most fundamental law of physics. Entropy has sometimes been called 'time's arrow'.

Entry charge: (see usage charge!)

Ethanol: Ethyl alcohol/grain alcohol from which a fuel can be blended. The alcohol itself can be made from any sugar or starch.

Ex-Ante and Ex-Post: Before an event; and after (the same event).

Externality: An action by either a producer or consumer that affects other firms or producers, yet is not accounted for in the market price. External economies are positive; external diseconomies are negative.

Faustian dilemma: A situation characterized by advantages in the short run that are purchased at the cost of disadvantages in the long run.

First law of thermodynamics: The equivalence of heat and work. Put another way, the total energy in the universe is constant: energy cannot be created or destroyed. At the same time, however, it can be transformed from the useful to the non-useful type (which is what entropy is all about).

Flue-gas desulfurization (i.e. the scrubbing of stack gases): Injecting materials such as limestone and water into a smokestoack to react with the sulphur in exhaust gases. A problem here is that scrubbing stack gases can oriduce sulfuric acid that must be collected and efficiently disposed of.

Fluidized-bed combustion: This involves crushed coal being fed into a boiler compartment containing small particles of limestone placed on top of a metal grid. When hot air is forced up through the grid the coal and limestone mixture begins to float and expand, and eventually the solid particles commence displaying the properties of a boiling liquid - the origin of the expression 'fluidized bed'. A large part of the pollutants in coal are removed before the flue gases leave the boiler, and in theory even high sulphur coal can be successfully treated by this process.

Forward contract: A contract involving the sale or purchase of an asset at some point in the future. The price can be determined in the present, or it can be related to the spot price at the date when the transaction is terminated, or just before that time.

Fuel cells: Fuel cells basically involve electrolysis in reverse: instead of electricity transforming water into hydrogen and oxygen, hydrogen and oxygen are transformed into water and electricity. Think of the way that batteries function. Although gaseous hydrogen is often mentioned as the most promising fuel (or input) in this process, such things as methanol and propane could be used, among others. Fuel cells can also operate on renewable resources, such as decaying refuse or vegetable matter, and in theory extremely high efficiencies can be achieved. In many ways this technology seems ideal, but up to now it has been a disappointment, cost-wise.

Fusion: Fusion is the integration of two ligher nuclei into one heavier nucleus (while fission is the disintegration of a heavy atomic nucleus into two lighter nuclei). The idea is to fuse the heavy isotopes of hydrogen (e.g. tritium or deuterium atoms) into helium atoms at a controlled rate, which means the availability of the same quantities of energy that are liberated in large thermonuclear explosions. This technology will not be available in the near future, although if I am mistaken, it will be the most important technological breakthrough of the 21st century.

Gas-graphite reactor: A reactor in which the fuel – conventionally - is natural uranium in metallic form (instead of enriched uranium).. The moderator is carbon, and the coolant is carbon dioxide gas under pressure. The well known 'Magnox' class of reactors is a gas-graphite reactor. The same is true of the Advanced Gas Reactor (AGR), although this equipment uses enriched uranium as fuel.

Gas turbine generator: Equipment where fuel is mixed with air compressed in a compressor stage, then burned to generate mechanical power (via a movable part), which is further converted to electrical power via a generator.

Geothermal energy: Roughly, underground steam that can be extracted and used for driving electric turbines, as well as e.g. heating. Geothermal energy does not produce atmospheric particulate pollution, but it has, at times, been associated with major emissions of hydrogen sulfide, sulphuric acid, etc. So much so that in Japan. In 1974, a permit for the sinking of 14 geothermal wells was revoked after 6 wells had been sunk.

Greenhouse effect: The warming of the earth's surface due to the 'trapping' of heat from the burning of fossil fuels. What happens is that the carbon dioxide concentration increases, which in turn brings about atmospheric conditions that promote an increase in temperature.

Heat rate: The number of e.g. British Thermal Units (Btu) needed to obtain one kWh in an actual system. In a perfect system this is 3,412 Btu.

Heavy Oil: Oil that is similar to conventional crude oil, but extremely viscous, and as a result relatively expensive to extract and process. Found mostly in Canada and the United States, although the largest quantities may be in Venezuela.

Heavy water reactor: (See light water reactor also). The best known of these reactors is the Canadian CANDU. The moderator (and in the CANDU the coolant) is heavy water, usually an oxide of deuterium. One of the features of this reactor is that it can use natural uranium (as compared to enriched uranium). See also 'moderator' below.

Hedging: In this book 'price insurance' using options, futures, and swaps. It can also take place using e.g. long term contracts.

Hydroelectricity: Electricity produced from flowing water. The energy produced is generally considered 'clean', but thee can be negative environmental effects. The conversion of a rapidly flowing river into a much slower moving one can mean heavy silting – where silt is sediment/sludge/mud etc – and can also mean a higher incidence of water borne diseases in the vicinity of the dam. Another often mentioned environmental cost is the flooding of wild-life santuaries.

Hydrogen economy: Using plentiful electricity – if it is plentiful – to obtain hydrogen from such things as electrolysing water. The hydrogen would then be used as a vehicle fuel, burned in a furnace for space heating applications, or used as a fuel for fuel cells in applications where it was undesirable to degrade chemical energy into heat. Ostensibly, hydrogen would become the energy basis of the economy.

In-the-money-option: An option that could be exercised profitably immediately.

Inelastic demand: When an x% change in the price of a good causes a change in demand – usually in the opposite direction – that is that is less than x%. Elastic demand is just the opposite.

Investment cost: The expenditure occurred now in the expectation of future rewards.

Joint venture: An economic undertaking by two wholly separate economic identities. For example, the construction of an electricity generating facility in India by and Indian and a US firm.

Light Water Reactor (LWR): Most of the world's power reactors are light water reactors. They are characterized by their use of light (i.e. ordinary) water as both moderator – which slows down neutrons – and coolant (to absorb the heat generated). These reactors come in two main types - pressurised water reactors (PWR) and boiling water reactors. The fuel is enriched uranium: 97% uranium-238, and (approximately) 3% uranium-235.

Magnox reactor: A reactor developed in the UK that is gas cooled, and operated on natural (i.e. non-enriched) uranium.

Market power: The ability of a producer to profitably affect price.

Merchant plant: A power plant from which all output is due to be sold spot. (But ex-post other arrangements are possible.)

Methane: The principal constituent of natural gas. Now considered to be an important contributor to the 'greenhouse effect'.

Methanol: Sometimes called synthetic natural gas, but it can be reduced to a liquid by refrigeration. It can be produced from e.g. coal, or the dry distillation of wood, and in the latter case it means that it is a kind of wood alcohol. Methanol is low on heat content, but the new transport technology seems to be moving in the direction of methanol and fuel cells, with methanol being the primary input for fuel cells.

Moderator: A light material (that is mixed with the fuel in a reactor), and whose function is to slow down the neutrons in certain types of reactors. This 'slowing down' produces stable chain reactions, in that it raises the probability that one neutron will cause fission in another. But reactors can be designed that are based on fission caused by fast (high energy) neutrons.

Moral hazard: This is a situation that can arise when an individual who is insured can control, to some extent, outcomes – i.e. states of the world – that are relevant to the insurance contract or arrangement. For example, a person with full insurance coverage for his automobile might not be careful about whether he locks the door of his vehicle.

Nash equilibrium: Behavior – in a conflict or non-cooperative situation – where each player is doing the best it can, given the actions of all other players. In the prisoners' dilemma game, the Nash equilibrium involves smaller playoffs to the players than a cooperative solution.

Natural monopoly: When it is cheaper for one firm to produce a good or service at the output levels corresponding to 'likely' demand than it is for two or more firms to do so.

Non-systematic risk: Risk that can be eliminated by holding a well diversified portfolio.

Notional: In finance this term generally means the amount or value that underlies a specific transaction, regardless of e.g. the structure of payments. In discussing a swap, notional value means the value that is relevant to the transaction.

Oil shale: Oil shales have been formed geologically by the incorporation of organic matter into sedimentary rock. It does not contain petroleum in any form, and what it does contain can be thought of as a substitute crude oil. A great deal of water is necessary to produce the substitute crude, and there are large environmental costs associated with this production. On the other hand, there is a huge amount of it available. The largest reserves seem to be in Brazil, Central Africa, the Former Soviet Union (FSU), and the US.

Opportunity cost: The cost associated with opportunities that are foregone by not using resources in their highest value use, which can be taken as the most feasible alternative. For instance, the alternative to a risky investment in which some of the possible outcomes might be less than the yield on a government security, is investing in the government security. The expression that is most often used here is 'next best alternative'.

Option: A financial contract allowing a transactor to buy (a call option) or to sell (a put option) an asset at a given price (the exercise price) during a specified period of time. The price of this option is called the premium.

Option writer: The seller of an option.

Out-of-the-money-option: An option that cannot be exercised profitably at a given point in time.

Peak Shaving: The injection of supplemental supplies of natural gas from e.g. underground storage or LNG facilities into a pipeline system during periods of maximum demand. With electricity, cycling the use of equipment (e.g. air conditioners) over short periods of time in order to help shave demand.

Petroleum: Oil + gas.

Polluter pays principle: The polluter should bear the costs of measures required to achieve a desired environmental standard; and these costs should be reflected in the prices of the goods and services provided by the polluter.

Portfolio: A collection of different assets held by a transactor.

Power: The rate at which work is being done, or can be done. Or, what amounts to the same thing, the amount of energy converted per uit of time. For example, a machine might be able to handle an indefinite amount of energy, but only at a finite rate. In the case of a light bulb, for example, it might be perfected so that it could be on an unlimited amount of time, but the illumination provided is determined by its power rating (in watts).

Primary energy: Energy from the direct burning of fossil fuels, as well as hydroelectricity and uranium based electricity.

Rate-of-return regulation: Setting a price that gives a firm – usually a utility or a monopoly – a competitive return on its assets.

Refining: A refinery is an installation in which various products are produced – or 'refined' – from crude oil. The main functions of a refinery are distillation (i.e. vaporization of the various components of crude oil), and 'cracking', which involves separating lighter from heavier compounds. Refineries produce gasoline, kerosine, heating oil, etc.

Reservation price: the maximum amount that a consumer is willing to pay for a good.

Risk aversion: Not being willing to make a fair bet – which is not the same thing as avoiding all risk..

Risk-free asset: An asset free of risk of default – that is, having the same value in all states of the world. The yield on such an asset is called the risk-free return. One example might be a government bond whose yield was indexed to compensate for inflation.

Risk neutrality: Indifference between earning a certain income, and a risky income with the same expected value as the certain return.

Risk premium: The amount that a risk averse individual will pay to avoid taking a risk. For example, the additional interest in excess of the risk-free rate that an investor must receive to compensate for risk.

Rounding, or 'rounding off': Approximating. For instance, rounding 2.81716 to 2.82.

Salvage value: The amount for which a physical asset can be sold at the end of its useful life.

Second law of thermodynamics: Heat cannot of itself pass from a colder to a hotter body.

Secondary energy: Energy obtained by the conversion (e.g. burning) of primary fuels (e.g. fossil fuels) to other forms of energy (e.g. electricity), as well as energy obtained from such things as 'town gas' and coke.

Speculator: Someone who attempts to earn profits by predicting the future state of the market (i.e. future supply and/or demand).

Stack gas scrubbing: Removing some or all sulphur dioxide from stack gases. The problem here is that the scrubbing produces sulphuric acid which, unless collected, may cause extensive environmental damage.

Stranded investment (or stranded costs): Investment, usually in a public utility, that has been undertaken, but which is not yet profitable because of the slow growth of demand, and which may never become profitable because the change in the regulatory background results in increased competition, etc.

Strip mining: Mining – usually the mining of coal – in which the overburden is removed (stripped), and then the coal is removed from an open trench. The last stage involves drilling and blasting, and then loading the coal into e.g. trucks. There can be large environmental costs associated with strip mining, and in particular the cost of restoring land in order to counter visual disamenities. Strip mining is sometimes called open-cast mining.

Sunk cost: An expenditure that has been made and cannot be recovered, even in the long run.

Synthetic crude: Oil produced from e.g. coal.

Tar sands: Somewhat misnamed source of oil. Tar sands are sandstones in which the pore spaces are wholly or partially filled by congealed oil. The Athabasca tar sands of Canada are the best known producers of this resource.

Term structure of oil prices: The profile of oil prices over time – generally rising, falling, or flat, although some combination of these are possible. Future prices can be represented by forward prices or by 'futures' prices.

Time-of-day-pricing: Charging more for an output (e.g. energy) consumed during peaking hours.

Time value of money: The principle that a given sum of money has a greater value today than if the same sum was received in the future.

Tragedy of the commons: The elimination of (mostly social) gains due to the overuse of common property.

268

Two-part tariff: A kind of pricing in which purchasers of a good or service are charged both an entry fee and a usage fee.

Usage charge: Sometimes called an intry or customer charge or fee. For instance, a charge might have to be paid to enter an amusement part, and after that there are separate charges for the various activities. With electricity, there might be a power charge (for using the system), as well as an energy charge – or charge per kilowatt hour.

User cost: A cost associated with a depletable resource, which takes into consideration the fact that using a resource now eliminates the opportunity to use it in the future. Could be called the opportunity cost of depletion.

Utility: Usually a monopoly (or in some cases a firm in an oligopoly situation) that is egulated for the purpose of limited its profits. The pricing applied to these firms is often average cost pricing, where 'normal profit' is counted as a cost.

Winners' curse: The situation in which the winner in an auction or a similar activity discovers that the item is worth less than she thought.

A.4 REFERENCES

Adelman, M.A. (1962), The Supply and Price of Natural Gas. Oxford: Basil Blackwell.
" (1972), The World Petroleum Market. Baltimore, Maryland: Johns Hopkins University Press.

Alchian, A. and Allen, W. (1964), Exchange and Production Theory in Use. California: Belmont.

Amundsen, E. and Bergman, L. (1995), 'The deregulated electricity markets in Norway and Sweden', SNF Working Paper No. 2, Bergen, Norway.

Ang, B.W. (1995), 'Decomposition methodology in industrial energy demand analysis' Energy 20(11): 1081-1095

Angelier, J.P. (1994), Le Gaz Naturel. Paris: Economica.

Banks, F.E. (1972) 'An econometric model of the world tin economy: a comment', Econometrica 33(22): 24-28.
—— (1974), 'A note on some theoretical issues of resource depletion', Journal of Economic Theory, 5: 125-32.
—— (1976), 'Laying Hamlet's ghost in commodities', New Scientist, August.
—— (1984), 'A note on optimal systems planning', The Energy Journal, 5(4): 79-80.
—— (1987), The Political Economy of Natural Gas. Beckenham, UK: Croom-Helm.
—— (1991), 'Paper oil, real oil, and the price of oil' Energy Policy, 6:1-15.
—— (1994), 'Sweden's electricity reform', The IAEE Newsletter, Summer.
—— (1999), 'Another side of European natural gas deregulation', Hydrocarbon Asia, May-June.

Beck, R.J. (1996), 'Strong demand seen for oil and gas in 1997-99', Oil and Gas Journal, 47: 169-174.

Berg, E. (1995), 'Miljøvirkninger av Norsk gass salg – en tilbudssideanalyse', Sosial Økonomen, 11: 18-25.

Bernard, J-T., Bolduc, D., Gingras, Y. and Rilstone, P., 'La demande d'électricité

Quebec', Energy Studies Review, 5(1): 28-37.

Black, F. and Scholes, M. (1973), 'The pricing of options and corporate liabilities', Journal of Political Economy, 81: 133-147.

Boxall, J. (1991), Pollution in the Metropolitan and Urban Environment. Hongkong Institution of Engineers: Hongkong.

Caffara, C. (1990), 'Petroleum stocks, role perceptions and reality', Oxford Energy Forum, 1: 40-43.

Calabre, S. (1992), 'Les marchés a terme: l'experience du petrole', Revue de L'Energie, 437: 99-112.

Carr, D.E. (1978) Energy and the Earth Machine. London: Abacus.

Chenery, H.B. (1949), 'Engineering production functions', Quarterly Journal of Economics, 63: 507-31.

—— (1952) 'Overcapacity and the acceleration principle', Econometrica, 20: 1-28.

Cleveland, C. and Kaufman, R. (1993), 'Forecasting ultimate oil recovery and the rate Of production', The Energy Journal, 12: 201-21.

Constanty, H. (1995), 'Nucleaire: le grand trouble', L'Expansion, 417: 68-73.

Crew, M.A. and Kleindorfer, P. (1979) Public Utility Economics. London: MacMillan.

Criqui, P. (1991), 'Apre la crise du Golfe, le trosieme chock petrolier reste a venir', Revue de L'energie, 437: 88-98.

Davis, G.A. (1996), 'Option premiums in mineral asset pricing', Land Economics, 72(2): 167-186.

Dixit, A.K. amd Pindyck, R.S. (1993) Investment Under Uncertainty. Princeton: Princeton University Press.

Drollas, L.P. (1986), Oil supplies and the world oil market in the long run. (Stencil).

Dunn, J. A. (1999), 'The politics of automobility', Brookings Review, 17(1): 40-43.

Edgeworth, F.Y. (1881) Mathematical Psychics. London: Kegan Paul.

Ellerman, D. (1996), 'The competition between coal and natural gas', Resources Policy, 22(1): 33-42.

Fesharaki, F. (1996), 'The outlook for oil demand, supply, and trade in the Asia-Pacific region to 2005', IAEE Newsletter (Fall).

Flower, A. (1978), 'World oil production', Scientific American, 238(3): 41-49.

Finon, D. (1989), 'Uranium in the shadow of a world wide nuclear power crisis', Energy Studies Review, 1: 28-44.

Funk, C. (1991), Wettbewerb im erdgassmarkt: welche ordnungs-rahmen sind möglich. Stencil: University of Bergen.

Georgescu-Roegen, N. (1971) The Entopy Law and Economic Process. Cambridge: Harvard University Press.

Gerondeau, C. (1984) L'energie a revendre. Paris: Lattés.

Giret, V. (1996) 'Ici on démonte le socialism nucléaire', L'Expansion, 520: 50-52.

Gordon, R. (1978) Coal in the U.S. Energy Market. Lexington MA: D.C. Heath.

Granström, T. (1982) Elförsörjningen i Energidebattens Centrum. Stockholm: Liber.

Green, R.J. (1995), 'Reform of the electricity supply industry in the UK', Journal of Energy Literature, 2: 3-24.

Haas, R.,Zöchling, J., and Schipper, L. (1998), "The relevance of asymmetry issues for Residential oil and natural gas demand', The OPEC Review 32(2):113-145.

Hall, D.C. (1996), 'Geoeconomic time and global warming' , International Journal of Social Economics, 23: 64-87.

Hansen, U. (1998), 'Technological options for power generation', The Energy Journal, 19(2): 63-87.

Harlinger, H. (1975), 'Neue modelle für die zukunft der menscheit'. Discussion paper. IFO-Institut für Wirtschaftsforschung, Munich.

Hjalmarsson, L. (1986). Prissättningen på elenergi och elmarknaden's utveckling. (Stencil.) Stockholm: Vattenfall.

Hope, C. and Gaskell, P. (1985), 'The competitive price of oil', Energy Policy, 8:289-296.

Horsnell, P. (1997), 'Crude oil price data: an empiricist's guide', Journal of Energy Literature, 3(1): 29-40.

Hotelling, H. (1931), 'The economics of exhaustible resources', Journal of Political Economy, 39: 137-75.

Howell, D.G., Bird, K.J., and Gautier, D.L. (1993), 'Oil: when will it run out?', Earth, 2: 26-33.

Hull, J. (1995) Introduction to Futures and Options Markets. Englewood Cliffs, New Jersey: Prentice Hall.

Iledare, O. and Pulsipher, A.G. (1999), 'Trends in US oil production warn of trouble', World Oil. (February):108-41.

Jacquet, P. and Nicolas, F. (1991) Petrole: Crises, Marchés, Politiques. Paris: Dunod.

Joscow, P.L. and Jones, D.R. (1983), 'The simple economics of industrial cogeneration' The Energy Journal, 4(1): 1-22.

Kapitza, P. (1976), 'Physics and the energy problem', New Scientist, 11:96-103.

Khartukov, E.M. (1993), 'East Asia's energy security: a Russian perspective', PetroMin 25(5): 8-20.

Krautkraemer, J.A. (1998), 'Nonrenewable resource scarcity,' Journal of Economic Literature, 36: 2065-2107.

Knudsen, K. (1984), 'Den glemte årsak til oljeprisens mulige kollaps,'' Sosial Økonomen, 4: 21-23.

Laherrere, J. (1995), 'World oil reserves – which number to believe', OPEC Bulletin, 14: 9-13.

Leckie, G. (1998), 'Oil Reserves – approaching the half-way mark?', OPEC Bulletin, 10: 8-10.

Layard, P., and Walters, A.A. (1978) Microeconomic Theory, New York: McGraw-Hill.

Lohrenz, J. and Bailey, A.J. (1995), ' Evidence and results of present value maximization for gas and oil development projects', Proceedings, Society of Petroleum Engineers.

Lynch, M.C. (1999), 'Oil scarcity, energy security, and long term oil prices – lessons Learned and unlearned', IAEE Newsletter, (First Quarter).

MacFadyen, A.J. (1990), 'Inter-relationships between spot and sticky-price markets   for Crude oil', Energy Policy, 6:168-173.

MacKerron, G. (1992), 'Nuclear costs: why do they keep rising?' Energy Policy, July: 641-652.

Marks, R. and Swan, P. (1993), 'Exhaustibility and the reserve/production ratio', Economics Letters, 42: 1-4.

Marshalla, R.A. (1978), An Analysis of Cartelized Market Structure for Nonrenewable Resources. PhD dissertation, Stanford, California.

Martin, J-M. (1992), Economie et Politique de L'energie. Paris: Armand Colin.

Masseron, J. (1992), 'Impacts of the Gulf War and changes in Eastern Europe', The Energy Journal, 13(3):1-16.

Matsui, K. (1998), 'Global demand growth of power generation', The Energy Journal, 19(2): 93-107.

Miller, M.H. and Upton, C.W. (1974), Macroeconomics: A Neoclassical Introduction.

Homewood, Illinois: Richard D. Irwin.

Murcier, A. and Boissonnat, J. (1982), 'Petrole: la grande illusion', L'expansion. 2: 15-23.

Nash, J.F. (1950), 'The bargaining problem', Econometrica, 22, 155-162.

Neumann, J. and Morgenstern, O. (1944) Theory of games and economic behavior. Princeton: Princeton University Press.

Newbery, D.W. (1995), 'Power markets and market power', Energy Journal, 16: 39-66.

Nordhaus, W.D. (1973), 'The allocation of energy resources', Brookings Papers on Economic Activity, 3: 529-70.

Owen, A.D. (1985) The Economics of Uranium. Praeger: New York.

Percebois, J. (1989) Economie de L'energie. Paris: Economica.

Pickles, E. (1994), 'Convenience yield, volatility, and the equilibrium price of oil', (Stencil).

Pindyck, R.S. and Rubinfeld, D.L. (1992) Microeconomics. New York: MacMillan.

Plourde, A. and Watkins, G.C. (1999), 'Changing relationships between crude oil and Natural gas prices', Paper for the 22nd IAEE Annual Meeting, Rome: 9-12 June.

Poundstone, W. (1992) Prisoners Dilemma. Oxford: Oxford University Press.

Radetzki, M. (1995) Tjugo År Efter Oljekrisen. Stockholm: SNS Förlag.

Rayment, P.B.W. (1980), 'Petrol prices, conservation, and macroeconomic policy', Economic Commission for Europe, Geneva (Switzerland).

Roncaglia, A. (1994), The effects of carbon taxes in different approaches to economic Theory', The Manchester School, 62(4): 438-444.

Rees, R. (1984) Public Enterprise Economics. London: Weidenfeld and Nicolson.

Ross, P. (1996), 'Reenter yellowcake', Financial Times Energy Economist, 73: 17-20.

Rowse, J. (1986), 'Allocation of Canadian natural gas to domestic and export markets', Canadian Journal of Economics, 19:417-442.

Salant, S.W. (1995), 'The economics of natural resource extraction: a primer for Development economists', World Bank Research Observer, 10: 93-11.

Salameh, M.G. (1999), 'Technology, depletion, and the myth of the reserve-production ratio', OPEC Review', 23(2): 114-125.

Saunders, H.D. (1984), 'On the inevitable return of higher oil prices', Energy Policy, September: 310-320.

Scarf, H. (1994), 'The allocation of resources in the presence of indivisibilities', Journal of Economic Perspectives, 7: 111-128.

Schultz, W. (1984), 'Die langfristige kostenenwicklung für steinkohle am weltmarkt', Zeitschrift für Energiewirtschaft, 1: 24-38.

Sharma, D. and Bartels, R. (1997), 'Distributed electricity genertion in competitive energy markets: a case study in Australia', The Energy Journal (Special Issue).

Siebert, H. (1979) 'Erschopfbare ressourcen', Wirtschaftsdienst, 10: 1-7.

Slesser, M. (1978) Energy in the Economy. London: MacMillan.

Smith, J.L. (1995), 'On the cost of lost production from Russian oil fields', The Energy Journal, 16(2): 25-57.

Smith, V. (1961) Investment and Production. Cambridge MA: Harvard University Press.

Solow, R.M. (1974), 'The economics of resources or the resources of economics', American Economic Review, 64: 1-14.

Späth, F. (1983) 'Die preisbildung für Erdgas', Zeitschrift fur Energiewirtschaft 3: 99-101.

272

Stern, J.P. (1984) Natural Gas Trade in Europe. London: Heinemann.

Stevens, P. (1995), 'The determination of oil prices: 1945-95', Energy Policy, 23: 10-23.

Stoppard, M. (1996), 'A new order for gas in Europe', Oxford Institute for Energy Studies, (Oxford UK).

Teece, D.J. (1990), 'Natural gas in Germany and the United States', The Energy Journal, 11: 1-18.

Teitelbaum, R. (1995), 'Your last big play in oil', Fortune, October 30.

Tempest, P. (1996), 'Defining and overcoming risk – some global and Middle East factors', IAEE Newsletter, 5: 9-10.

Thunell, J. (1979) Kol, Olja, Kärnkraft – en Jämförelse. Stockholm: Ingenjörsförlagen.

Ulph, A.M. and Folie, G.M. (1978) 'Gains and losses to producers from the cartelizationOf exhaustible resources', Working Paper No. 369, Center for Resources and the Environment, Canberra: The Australian National University.

Valle, A.P.D. (1988), 'Short-run versus long-run marginal cost pricing', Energy Economics,10: 222-235.

Varian, H.R. (1993) Intermediate Microeconomics. New York: W.W. Norton.

Verleger, P.K. (1993), 'Adjusting to volatile energy prices', Discussion paper, Institute for International Economics, Washington.

Vollans, G.E. (1999), Real-time retail pricing of electricity. Paper presented at the 22nd IAEE Annual International Conference, Rome, 9-12 June.

Wallin, M. (1996), Handel i kraft. Nationalekonomiska Institutionen, Uppsala. (Stencil).

Watkins, G.C. (1992), 'The Hotelling principle: autobahn or cul-de-sac', The Energy Journal, 13(1): 1-24.

Weiner, R. (1996), 'Middle East crude oil pricing and risk management in the 1990s', Journal of Energy Finance and Development, 1: 21-49.

Wilkinson, J.W. (1983), 'The supply, demand, and average price of natural gas under Free-market conditions', The Energy Journal, 4(1): 99-123.

Wirl, F. (1994), 'The world oil market after the Iraq-Kuwait crisis', Energy Sources, 16:75-88.

Yergin, D. (1991) The Prize. New York: Simon and Schuster.

Zimmerman, M.B. (1981) The U.S. Coal Industry. Cambridge MA: MIT Press.

Index

274

supply of storage, 67
Sweden, 4
swing producers, 116
swing supplier, 93
'synoil' and 'syngas', 9

Tarif Vert, 195
Tartakover, Savielly, 249
Teece, David, 96
Tempest, Paul, 48
The Economist, 2
Time (magazine), 2
Tokyo Power Company (Tepco), 180
Tracy, Spencer, 50
transportation economics (coal), 135
Trinidad and Tobago, 81

United States Environmental Protection
Agency (EPA), 129
Upton, Charles, 235
uranium yellowcake, 205
US Energy Information Administration
(EIA), 43
utilities (public), 18

value of known reserves, 235
Venezuela, 73
Verleger, Philip, 144
Vind, Karl, 236
Walls, W. David, 240
Walters, Alan, 99
Waterpower, 18
watt, 14

Venezuela, 73
Verleger, Philip, 144
Villagrasa, Delia, 8

West Texas Intermediate, 144
wind power, 19, 207
Winner's Curse, 240, 242
World Climate Conference, 7
World Coal Institute, 113
World Energy Council, 3, 45
World population, 200
World Trade Building, 222

Yergin, Daniel, 47

Åre (ski resort), 32, 60, 244